FOOD, AGRICULTURE AND SOCIAL CHANGE

In recent years, food studies scholarship has tended to focus on a number of increasingly abstract, largely unquestioned concepts with regard to how capital, markets and states organize and operate. This has led to a gulf between public policy and people's realities with food as experienced in homes and on the streets. Through grounded case studies in seven Latin American countries, this book explores how development and social change in food and agriculture are fundamentally experiential, contingent and unpredictable.

In viewing development in food as a socio-political-material experience, the authors find new objects, intersubjectivities and associations. These reveal a multiplicity of processes, effects and affects largely absent in current academic literature and public policy debates. In their attention to the contingency and creativity found in households, neighborhoods and social networks, as well as at the borders of human–nonhuman experience, the book explores how people diversely meet their food needs and passions while confronting the region's most pressing social, health and environmental concerns.

Stephen Sherwood is Lecturer and Researcher in Knowledge, Technology and Innovation at Wageningen University, the Netherlands.

Alberto Arce is Associate Professor of Sociology of Development and Change at Wageningen University, the Netherlands, and Associate Professor at the Faculty of Agronomy at the Universidad de Chile.

Myriam Paredes is Professor of Rural Territorial Development at the Facultad Latinoamericana de Ciencias Sociales (FLACSO), Ecuador.

Other books in the Earthscan Food and Agriculture Series

For further details please visit the series page on the Routledge website:
www.routledge.com/books/series/ECEFA

Climate Change and Agricultural Development
Improving resilience through climate smart agriculture, agroecology and conservation
Edited by Udaya Sekhar Nagothu

Forgotten Agricultural Heritage
Reconnecting food systems and sustainable development
Parviz Koohafkan and Miguel A. Altieri

International Law and Agroecological Husbandry
Building legal foundations for a new agriculture
John W. Head

Food Production and Nature Conservation
Conflicts and Solutions
Edited by Iain J. Gordon, Herbert H.T. Prins, and Geoff R. Squire

Sustainable Intensification in Smallholder Agriculture
An integrated systems research approach
Edited by Ingrid Öborn, Kwesi Atta-Krah, Michael Phillips, Richard Thomas, Bernard Vanlauwe, and Willemien Brooijmans

Environmental Justice and Farm Labor
Rebecca E. Berkey

Plantation Crops, Plunder and Power
Evolution and exploitation
James F. Hancock

Food Security, Agricultural Policies and Economic Growth
Long-term dynamics in the past, present and future
Niek Koning

Agriculture and Rural Development in a Globalizing World
Challenges and Opportunities
Edited by Prabhu Pingali and Gershon Feder

FOOD, AGRICULTURE AND SOCIAL CHANGE

The Everyday Vitality of Latin America

Edited by Stephen Sherwood, Alberto Arce and Myriam Paredes

Routledge
Taylor & Francis Group
LONDON AND NEW YORK

earthscan
from Routledge

First published 2017
by Routledge
2 Park Square, Milton Park, Abingdon, Oxon OX14 4RN

and by Routledge
711 Third Avenue, New York, NY 10017

Routledge is an imprint of the Taylor & Francis Group, an informa business

British Library Cataloguing-in-Publication Data
A catalogue record for this book is available from the British Library

Library of Congress Cataloging in Publication Data
A catalog record for this book has been requested.

ISBN: 978-1-138-21497-2 (hbk)
ISBN: 978-1-138-21498-9 (pbk)
ISBN: 978-1-315-44008-8 (ebk)

Typeset in Bembo
by FiSH Books Ltd, Enfield

"*Food, Agriculture and Social Change* forces us to rethink how we have conventionally thought about food and agriculture. Leaky embodiments. Relationality. Ontological multiplicity. Practice. Be prepared for a wild ride. The insights and hopeful critiques you'll be exposed to along the way make the journey a rewarding one indeed."

– Michael S. Carolan, Department of Sociology, Colorado State University, USA

"Challenging dominant patterns of socio-economic development in agriculture and food and proposing a critique of alternative views of production and consumption, the collection of essays included in *Food, Agriculture and Social Change* contains incisive and empirically-based analyses of the ways in which people define and carry out food production and consumption in Latin America. Stephen Sherwood, Alberto Arce and Myriam Paredes edited a book that, transcending established normative discourses and contesting abstract understandings of what agriculture and food 'ought to be,' demonstrates how agri-food is experienced and practiced in every-day life. The result is a dynamic and engaging edited volume that is a 'must read' for those interested in agriculture and food and socio-economic development in Latin America."

– Alessandro Bonanno, Texas State University System Regents' Professor and Distinguished Professor of Sociology, Sam Houston State University, Texas, USA

"Combining rich empirical material and provocative theoretical reflection, this book offers a fresh perspective on how food – which is simultaneously material, meaningful, relational and political – is both constituted through and constitutive of a vast array of practices in diverse social locations. By emphasizing the transformative power of people's agency through everyday practices, this book offers a strong counterpoint to technocratic presumptions common in global development institutions that food and agricultural systems can best be changed through policy initiatives and technological innovations."

– Todd Crane, Ecological Anthropologist and Senior Scientist, International Livestock Research Institute, Nairobi, Kenya

"*Food, Agriculture and Social Change* accounts for the everyday, vital assemblages of food, people and processes to help us not just construct a post-disciplinary social theory of food but this magisterial book places this accounting at the veritable centre of the re-production of our collective food futures. This volume provides an invaluable exploration of the processes by which food is more-than-food in its multiplicities, relationalities and intersectionalities and how this fluidity creates a rhizomatic politics that cuts across space, place and territory. *Food, Agriculture and Social Change* opens up a novel and very much needed window to the everyday politics of food and its socio-ecological vitalitisms in ways that will challenge the field of food studies for years to come."

– Michael Goodman, Professor of Environment and Development/Human Geography, University of Reading, UK

CONTENTS

CONTRIBUTORS

Alberto Arce is an Associate Professor at Wageningen University and has held academic positions at universities in England and Latin America. He has written on the changing relationships between the state, rural producers, and food globalization, and on the sociology and anthropology of an actor-oriented perspective to contemporary and diverse modernities. His current research interests include assemblage theory, social ontology, new materialisms, political ecology, and posthumanism in an emergent cosmopolitics of imperceptible worlds. Contact: Sociology of Development and Change, Wageningen University, the Netherlands; alberto.arce@wur.nl

Peter Berti is Nutrition Advisor and Deputy Director for HealthBridge, a Canadian NGO working in global health. He carries out research and leads programs on nutrition and food systems, with a focus in the Andes. Of particular interest is the relationship between agriculture production and diet of farming families. Contact: HealthBridge, Ottawa, Canada; pberti@healthbridge.ca

Gustavo Blanco studies the interfaces between development and environmental policies and people's everyday life. He is particularly interested in the creation of techno-political objects, the emergence of new subjects, and the co-production of knowledge in rural development processes and scientific initiatives related to global environmental change in southern Chile. He leads the research program "The Social Life of Energy". Contact: Instituto de Historia y Ciencias Sociales, Universidad Austral de Chile, Valdivia, Chile; gblanco@uach.cl

Flávia Charão-Marques is an Associate Professor at the Federal University of Rio Grande do Sul (UFRGS). Her research concentrates on sociotechnical heterogeneity in agriculture, food, and local health practices, with special attention to innovative processes and social change emerging from agroecology, rural women and food movements. Contact: Programa de Pós-Graduação em Desenvolvimento Rural/UFRGS, Porto Alegre, Brazil; flavia.marques@ufrgs.br

Maruja Cortés Belmar studies territorial development, agrifood heritage as an element for local development, core studies, agri-food policy and sustainable rural development (NEPAD) in the Department of Agricultural Economics, Faculty of Agricultural Sciences, University of Chile; marcortes@uchile.cl

Charlynne Curiel is interested in the emergence of organic food production and consumption and ideals of sustainability among growers, activists, and consumers. Through ethnographic work she explores the practices and relations between people and nature involved in the constitution of new subjectivities, associations, and communities. She participates in some of these organizations in the promotion of bio-intensive and organic food production and consumption in the city of Oaxaca. Contact: Instituto de Investigaciones Sociológicas, Universidad Autónoma Benito Juárez de Oaxaca, Mexico; curiel.iis.uabjo@gmail.com

Ana Deaconu is a PhD candidate in nutrition at the University of Montreal, where she hopes to better understand the integral relationship between the body and its food in context of its place, in order to explore the constitution of health. Focused on the Andes, she is currently interested in rural nutrition, food- and agro-biodiversity, health-promoting food movements, and the embodiment of food. Contact: Nutrition Program, Faculty of Medicine, University of Montreal, Canada; ana.deaconu@umontreal.ca

F. Vladimir Díaz Copado studies lifestyle mobility (so-called lifestyle migration) and its impact on Mexican host communities. He focuses on the interaction between foreign lifestyle movers, original residents of lifestyle destinations, and other local actors involved in shaping the social life and physicality of these locations. He is currently a postdoctoral researcher at the Institut für Ethnologie, Ludwig Maximilians Universität, Munich, Germany; vladimir.diaz@ethnologie.lmu.de

Jessica Duncan researches relationships between global governance mechanisms, food provisioning, the environment, and the actors that interact across these spaces. More specifically, she is interested in better understanding ways in which non-state actors participate in supra-national policy making processes, and analyzing how the resulting policies are implemented, shaped, challenged and resisted in localized settings. Contact: Rural Sociology Group, School of Social Sciences, Wageningen University, the Netherlands; jessica.duncan@wur.nl

Eleanor Fisher studies agrifood issues and livelihood change, with a body of recent research on rural employment dynamics related to smallholder agriculture and artisanal mining. Contact: School of Agriculture, Policy and Development, University of Reading, United Kingdom; e.fisher@reading.ac.uk

Christine Franke investigates the underlying relationships of people's practices through the lenses of food, agriculture, rural development and place. What triggers her interest is the need to understand the stories, emotions, desires, difficulties and concerns that guide people in their daily lives and that make them potentially powerful and creative actors. Contact: Rural Sociology, Wageningen University, the Netherlands; franke-christine@gmx.net

Jilles van Gastel works on the critical analysis of development policy and practice with special attention for new and renewed forms of organization including public–private partnerships and cooperatives. She is involved in the research program "The Social Life of Energy" for which she studies energy transition in Chilean agriculture using case-study methods. Furthermore, she teaches on various Master's programs. Contact: Instituto de Economía Agraria, Universidad Austral de Chile, Valdivia, Chile; jilles.vangastel@uach.cl

Joan Gross is Professor of Anthropology at Oregon State University where she helped start the Food in Culture and Social Justice program. Her current research focuses on food activism in Oregon and Ecuador and food and migration. Contact: Department of Anthropology, Oregon State University, Corvallis, Oregon; jgross@oregonstate.edu

Carla Guerrón Montero (PhD in Cultural Anthropology) is a cultural and applied anthropologist trained in the United States and Latin America, and specialized in the anthropology of tourism, the anthropology of food, and the African diaspora. Dr. Guerrón Montero is Associate Professor of Anthropology at the University of Delaware. She studies the multiple representations of identity among marginalized populations in modern Latin American and Caribbean nation-states, particularly through tourism and cuisine (Brazil, Ecuador, Grenada, and Panama). Contact: Department of Anthropology, University of Delaware, Newark, Delaware; cguerron@udel.edu

Michaela Hammer studies home healthcare and food practices of rural families, focusing on the dynamic interfaces among traditional and modern knowledges. She currently works to alleviate food insecurity in the United States by engaging diverse populations in organic food production at a small educational farm run by a non-profit food bank. Contact: Food for Lane County Youth Farm, Eugene, Oregon; michaela.hammer@gmail.com

Andrés Lagarrigue studies bio-cultural diversity in rural contexts with particular attention to agroecological practices and knowledge among peasant and indigenous communities. He is also interested in the study of social and cooperative economy and processes of territorial development. He participates in applied researches and is involved in several actions of the GIAHS (Globally Important Agricultural Heritage Systems) initiative of FAO at local level in Chiloé Island, Chile; lagarrigue@gmail.com

Horacio Narváez-Mena is a PhD candidate in Knowledge, Technology and Innovation. His research focuses on positive deviance in "water for food" practices, aimed at explaining knowledge co-production as expressed through heterogeneity in farming practice. Contact: Knowledge, Technology and Innovation, Wageningen University, the Netherlands; horacio.narvaezmena@wur.nl

Andrew Ofstehage is a PhD candidate in Anthropology. Before working on U.S. migrant farmers in northern Brazil, he studied quinoa commodification in Bolivia. Contact: Department of Anthropology, University of North Carolina at Chapel Hill; aofste@live.unc.edu

Daniela Oliveira is an independent consultant. She has experience with farmer's cooperatives, non-governmental organizations and government agencies on issues related to family farming, rural development, and agroecology. Her main research interest is rural innovation and social action. Currently Daniela is Project Director, City of Antônio Prado, Serra Region. Contact: Prefeitura Municipal de Antônio Prado, Rio Grande do Sul, Brazil; danioliveira02@yahoo.com.br

Myriam Paredes studies endogenous development of agriculture in the Andes. Through research on farming styles and policy processes, she explores the social heterogeneity embedded in agriculture and food practice. She is actively involved in social movements aimed at redefining and strengthening rural–urban linkages for greater human health, production, and sustainability. Contact: Desarrollo Rural Territorial, Facultad Latinoamericana de Ciencias Sociales (FLACSO), Quito, Ecuador; mcparedes@flacso.org.ec

Oscar F. Reyna is a PhD candidate who researches socio-environmental conflict at the confluence of mining, heritage, sacred natural sites and extractive industries. Conceptually, he explores the intersections of Posthumanism, New Materialisms and Multi-species ethnography. Oscar's main interest is to reveal the more-than-human territorial vitalities developing in places threatened by contemporary extractivism. Contact: Sociology of Development and Change, Wageningen University, the Netherlands; oscarfiles@hotmail.com

Claudia Job Schmitt is an Associate Professor at Rural Federal University of Rio de Janeiro (UFRRJ). She is member of the Observatory of Public Policies for Agriculture (OPPA), and her main research topics are agroecology, public policy, technologies, networks, social movements, food security and sovereignty and local markets. Contact: Departamento de Desenvolvimento Agricultura e Sociedade/Universidade Federal Rural do Rio de Janeiro, Rio de Janeiro, Brazil; claudia.js21@gmail.com

Stephen Sherwood studies the rising uncertainties associated with modernization in agriculture and food, concentrating on the institutionalization of unwanted, self-harmful and self-destructive activity around pesticides, obesity, and climate change. His research devotes special attention to people's self-organization and implications for science and development policy. Steve also operates a family-run organic farm in Ecuador, where he is involved in national and regional rural people's movements. Contact: Knowledge, Technology, and Innovation, Wageningen University, the Netherlands; stephen.sherwood@wur.nl

Paola Silva studies agricultural sustainability in Mediterranean annual crops. This line of research is divided into: (1) characterization and selection of adapted genotypes, where terminal drought is the main characteristic and (2) improvement of the agronomic efficiency of these ecosystems. Contact: Laboratorio de Relación Suelo-Agua-Planta, Departamento de Producción Agrícola, Facultad de Ciencias Agronómicas, Universidad de Chile, Santiago; psilva@uchile.cl

Camilo Torres is a PhD candidate at Wageningen University and Associate Professor in Economics at the Universidad Jorge Tadeo Lozano. He has worked for ten years with indigenous and rural communities in the Amazon of Colombia and Brazil. Camilo's research addresses local economic distortions, human ecology, and the differences between socio-environmental transactions and the new institutional economics. Contact: Facultad de Ciencias Económicas-Administrativas, Universidad Jorge Tadeo Lozano, Bogotá, Colombia; camilo.torres@utadeo.edu.co

Nicolas Vasconcellos is a PhD candidate who studies food sovereignty in Ecuador. In researching the relation between farming practices and policy applications, he explores social interactions related to peasant farming. He focuses especially on agricultural policy and implications on food production. Contact: Rural Sociology, Wageningen University, the Netherlands; nicolas.vasconcellosfernandez@wur.nl

Gerard Verschoor is Assistant Professor of Sociology at Wageningen University. He has studied human–environment relations in Burkina Faso, Colombia, Costa Rica, and Mexico. His current research interests focus on the ontological disjuncture that arises when modern institutions based on a nature–culture divide meet Indigenous peoples and the ensuing ontological politics. Contact: Sociology of Development and Change, Wageningen University, the Netherlands; gerard.verschoor@wur.nl

María Laura Viteri is interested in social relationships at locations of commercialization and consumption of food. She studies markets as collective devices that allow goods to be exchanged among people, who give different values to them. By focusing on social practices, she explores multiple processes of innovation as a journey of back-and-forth interactions between technologies and the people involved at diverse stages of development. Contact: Área de Economía y Sociología Rural/Programa Nacional de los Territorios, Instituto Nacional de Tecnología Agropecuaria (INTA), Buenos Aires, Argentina; viteri.maria@inta.gob.ar

ACKNOWLEDGMENTS

This ambitious, collective project to challenge critical social theory in agriculture and food in Latin America has followed a pathway with different phases of hope and frustration as well as inspiration. In the process, an eclectic collection of mind-bodies has converged around a common interest in strengthening empirical social research in the region through the application of flat ontology, practice theory, and ethnographic inquiry. On behalf of the editors, I would like to express my gratitude to the group of academics, food activists, and development practitioners who unselfishly contributed their experience and hard work in helping to consolidate a more critical social perspective in agrofood studies. Without their commitment and dedication, this book simply would not have been possible.

One-third of the 28 contributors are former students of Alberto Arce, and essentially everyone else, including Co-editor Myriam Paredes and myself, has been directly influenced by his work, be it through his writing, lectures and academic debates, traveling about a farm or in a market, or sitting down over conversation at a restaurant or bar. Those who know Alberto appreciate that he is tireless in his commitment to anthropology and the social sciences, as expressed in the attention he invests in his lectures, his students, and his colleagues. Sometimes Alberto is not just charitable with his time, but he also can be unsparing and even self-sacrificing in his willingness to defend a concept or idea. These qualities serve him well in his ability to continually push the frontiers of thought and imagination in development studies. While perhaps best known for his work with Norman Long and Terry Marsden on the actor-oriented perspective and its applications, Alberto continues to labor arduously in the expansion of his ideas, a tenacious exploration that recently has carried him to the realms of radical feminism, applied physics, and neurobiology to better understand the nuances of human (and human–nonhuman) living and being. In many ways, this book is a testament to that tireless effort and Alberto's boundless creativity and enthusiasm.

In our venture to produce something of value, we received much encouragement and support from a number of colleagues, in particular Eleanor Fisher at the University of Reading, United Kingdom. We also have benefited from ongoing discussions and debates with Severine van Bommel and Laurent Umans. In addition, we would like to acknowledge the support of Olivia Ange from the Marie Curie Intra European Fellowship, who was central in facilitating a workshop in the Netherlands to revitalize the sociology and anthropology of food. Working with Jessica Duncan, Clemens Driessen, Paul Hebinck, Maite Hernando-Arrese, Oscar Reyna, Vera Ribbens, Henk van den Belt, and Gerard Verschoor, Olivia played an important role in bringing together the Wageningen Ontology Group, where many of the ideas in this book were put forward, tested, and revised.

In April 2015, Mauro van Aken, Michela Badii, and Federica Riva organized a seminar at the University of Milan-Bicocca in Italy, in collaboration with the Fondazione Feltrinelli and the EXPO2015, during which a number of central insights emerged that ultimately inspired this project. It was at this seminar that initial ideas about food connections, disconnections, and reconnections surfaced to provoke fresh thinking on sustainable futures, the significance of experience and the paradoxes of pleasure. In addition, in August 2015 many of the contributors participated in a Special Seminar on "New Critical Social Perspective" at the Latin American Congress of Social Science, at FLACSO-Ecuador. Those activities led to a Spanish-language Dossier in *Revista Íconos*, entitled "Estudios críticos de la agricultura y la alimentación: perspectivas sobre el actor, la práctica y el territorio en América Latina" (No. 54, January 2016). We would like to express our appreciation for the support and encouragement of Jenny Ponton and Susana Wappenstein in the production of this earlier work. Lastly, I would like to thank colleagues at Misereor's Latin America Department, in particular Heinz Oelers, Malte Reshölft, and Heike Teufel, for involving me in their efforts to apply insights from a critical social perspective in their social change initiatives in the region.

During the final stages of production of this edited volume, we worked with a team of copy editors – Tom Edwards, Janneke Bruil, and Silvia Alunni – who were extremely generous with their time in helping us to (more or less) meet our submission deadline with Routledge. In particular we are grateful for the contributions of Janneke, who magnanimously edited just about every chapter, supplying not just grammatical corrections but also insightful qualitative comments that greatly contributed to improving the cohesion and clarity of the overall book. We also would like to thank Tim Hardwick, Ashley Wright, and their team at Earthscan/Routledge for their encouragement and attention to layout, printing, and distribution. Despite the heavy demands on his time as Senior Commissioning Editor, Tim was quick to recognize the potential of this project, and he provided knowledgeable and timely commentary that helped to improve and guide the progress of our ambitious proposal. We appreciate the support of each and every one of these people, but we take full responsibility for any remaining mistakes in the text.

Finally, we would like to acknowledge the influential financial support of the Dutch NWO/WOTRO through its Integrated Programme grant, "Positive

Deviance as a Catalyst for Sustainable Food Production and Nutrition in the Andes". This four-year research project financed the fieldwork of three chapters as well as a series of conferences and workshops in Ecuador, Portugal, and the Netherlands between 2012 and 2016 that enabled us to convene a collection of activist-scholars involved in agrofood studies to share and debate concepts, methods, and experiences. In addition, the grant financed the final production and printing of this book with Earthscan/Routledge. We would like to thank additional colleagues at EkoRural, FLACSO, and Wageningen University, in particular Ross Borja, Pedro Oyarzun, Martha Zurita, Cees Leeuwis, Vera Mentzel, and Inge Ruisch, for their competent and ample support to different aspects of the research and administration leading up to this publication.

Stephen Sherwood
Lead Editor

INTRODUCTION

Food embodiments, assemblages and intersubjectivities: ebbs and flows of critical food studies

Alberto Arce, Stephen Sherwood, and Myriam Paredes

Critical food studies

Economic subsidies and technological developments since the turn of the twentieth century, but in particular since the Second World War, have contributed to a revolution in the production, distribution, and consumption of food. To date, the field of agrofood studies largely analyzed developments by assessing the diffusion of technologies and change in state-based agriculture support policies. Compelling examples include studies of the emergence of cow's milk as a popular drink for humans in the United States (DuPuis, 2002), the industrialization of Dutch agriculture and the arrival of the "virtual farmer" (Van der Ploeg, 2003), and the rise of the international development industry and the institutionalization of technology transfer in the Americas (Flora and Flora, 1989). The nation-state, in both developed and developing countries, has strategically promoted highly intensive forms of food production and circulation, thereby making food an object and objective of economic development and productivity policies.

The understanding of food as an object of research and policy has led to the study of the formation of commodity complexes (Goodman, Sorj, and Wilkinson, 1987) as a result of a new international division of labor (Sanderson, 1985). A "critical approach" to food studies has aimed to demystify the world market through a historical comparison of major international food and agricultural regimes (Friedmann, 1993, Friedmann and McMichael, 1989; McMichael, 1994, Friedland *et al.*, 1991). Agrarian studies has pursued a range of political, economic, and cultural issues to address nuances of change in government and governance, in particular with regard to regulation (Marsden and Little, 1990; Lowe, Marsden, and Whatmore, 1994). By the end of the twentieth century, food studies became part of macro-explanations linking the social and political effects of national and international capital accumulation on employment, consumption norms, state regulation, legislation, and household livelihood and consumption practices (Goodman and Redclift, 1981).

These developments have tended to reify social institutions, displacing less formalized processes of social change, including the refashioning of food, beyond the responsibilities and control of people "on the streets" (for a review of this perspective, see Arce and Marsden, 1993 and for critical studies, Viteri and Arce, 2013; Fisher and Sheppard, 2012). As per this highly structural and functional analysis, there has been little or no room for actor agency and the possibility of multiple meanings of food. Nevertheless, this body of work has contributed important insights into food production and consumption as well as environmental, social, political, and cultural dimensions of global food processes. For example, studies of food as an expression of post-Fordism as well as of the emergence of non-state regulatory regimes in an era of growing internationalization and acceleration of commodity transfers called attention to changes in agricultural labor markets, the displacement of traditional farming and the related erosion of autonomy in peasant households (Bonanno, 1993), and the "nature and politics" of environmentalism and the state (Bonanno and Constance, 1996).

In this regard, Lynne Phillip's (2006) provocative commentary on food and globalization was a notable contribution. She proposed the repositioning of food studies itself as an object of research to explain "how food acquires mobility." Her review turned attention to three developments: the expansion of commodity relations beyond national borders, the rise of non-traditional export crops and foodstuffs, and the establishment of free trade agreements. She called attention to how multinational corporations aggressively made use of increased transnational mobility of food for their own profit and how multilateral agreements, in particular the General Agreement on Tariffs and Trade, financial institutions (World Bank and the International Monetary Fund), and other international organizations (the World Trade Organization) oriented the governance of food within and beyond individual nation-states towards consumption and private interests. In her conceptual unpacking of these seemingly abstract institutions, Phillips brought to life the neglected everyday actualities of food circulation.

Since then, the interrogation of contemporary flows of ideas, people and food has described and analyzed what happens to people who are "gated" to their localities, for example women trapped in family farming and in rural areas. Growing awareness of the geographic mobility of people's practices beyond state borders has contributed to thinking about food as a spatialization and semiotization of contemporary political economy (Lash and Urry, 1994).

A concern with food fragmentation, coordination, and reflexivity in a world of rising globalization has called attention to the need to "feed the global imaginary" and the tangible food materialities, socials, and naturals beyond a traditional geographic focus (see Massumi 1996), especially in reference to events in Latin America. Conceptual understanding of food as an object that encompasses producer and consumer positions and is tied to an unfolding world of inequalities reveals a new intersubjectivity in government, environment and politics. The ensuing rise of "food citizenship," expressed for example in cosmopolitan movements for agroecology and food sovereignty, calls attention to different global perspectives (Sherwood *et al.*, 2013). People who are organized in family, neighborhoods and social networks act in and act on emerging modes of existence (see Arce, 2009).

Contemporary critical food studies explores the significance of the migration of people, ideas and things in the reality of their food actualities. It explores how actors that are engaged in the re-internalization of global processes (re)script, for example, the family farm to the social life and market demands of the global social imaginary, forging potential contributions through agriculture and food practice. Such local and global dimensions of agrofood studies are not merely at conflict. Through encounters, relationships, and co-constructions, they also nourish one another in endlessly rich and recursive ways.

Although food scholarship has contributed to reconfiguring food as an object of global imagination, people (even food scholars themselves) have to eat and experience food. This happens in a diversity of ways: via microorganisms, taste, bodily sensation, and affective cognition. Food materialities are encountered in a multiplicity of physical forms, shaped by human design, natural vibrancy, and by the contingent moment. We see the global imaginary as an extension of the contemporary critical cognition of worldly materialities. Nevertheless, scholars have generally overlooked the dissolutions and re-assemblages of food – what we summarize here as the multiple objectivities and intersubjectivities of food. In this sense, the next step in agrofood studies, we argue, is to examine the re-localization of agriculture and food to the everyday lives of people – and within that, in the diverse spaces of food production, circulation and consumption.

Re-situating food in daily experience

Drawing on case material in the European Union, Maria Fonte (2010) provides insights into the empirical and policy "re-localization" of food. Following an exploration of academic controversies over local food, she concludes that alternative food ventures exist in the voids generated by the unfolding effects (and affects) of the standardization of agrofood industry. Fonte argues that the coexistence of alternative and conventional food is possible as long as competing food movements do not threaten the conventional agriculture and food industries. She identifies the principal obstacle of this re-localization and its competitive survival as the declining number of people in the rural space and the displacement of their knowledge. Fonte explains that this is the result of a science-dominated discourse that renders people's daily experience in agriculture and food to a technical event. For example, experts establish percentages of permissible emissions (e.g., fertilizer runoff, pesticide residues and more recently methane and carbon gases), thereby controlling access to "sustainability." Fonte argues, however, that the application of such normative knowledge to food blocks, rather than facilitates, sustainable rural development.

In our view, this normativity in the European Union as well as in similar examples of food re-localization in Latin America represents a situated, regional application of "ecological modernization," roughly understood as the state and regional institutionalization of norms for scientifically informed environmentalism and "best practice" (Mol, 2001). In its preoccupation with space and form and commitment to an alliance of expert and state-based regulation, public policy neglects or negates the rich intersubjectivities in the habitation of a territory. Fonte states that the local, understood as socio-spatial proximity, tends to emphasize the reconnection of producers and consumers through unique means of food circulation and exchange. In addition, Fonte speaks to a linguistic "localness," explained as the relationship between food and territory that rests in the specific conditions of experiencing a mode of existence and knowledge in a particular landscape. This material intersubjectivization of food, for example in creating new identities through regional labeling of cheese and wine, goes beyond the cognitive agency of human actors (see also Tsing, 2015). According to Fonte, socio-ecological interactions of locality generate a unique set of discourses that mobilize rural and urban actors, food producers, and consumers. As a means of addressing such interactions, critical food studies has become interested in examining processes of translation and the significance of vernacularization of material performances of food outside the normal neoliberal market conventions. While bringing together the material and the affective cognition of food, actors, and territory, these multiple discourses establish food as a highly creative and dynamic association involving mind, body, and the world.

Such discursive and material multiplicities, we suggest, mobilize subjects into subjectifications (e.g., the collective ennui of needing to recognize or defend one's food territory vis-à-vis competing interests), generating embodiments (a particular product, such as "champagne") within materiality (the emergence of price nuances, markets, and certification schemes), affectivity (determination of a form of feeling in romance) and cognitive-material food assemblages (struggles to determine the intangible properties of champagne and hence its "territory"). These are distributed and composed in physicochemical traits – for instance connections between mind and brain of individuals as well as in the manifestations of emerging corporealities. If a person loves wine and cheese, for example, that flavor is not just a matter of individuality, but also of taste, touch, smell, and collective memories – that is, social contagion (Gronow, 1997). In this associative orientation, human actors and nonhuman actants are bound in a multifarious world of social, cultural, natural, and material singularities – each an expression of the politics of life.

In their study of the food regions of Wales, Tuscany, and California, Morgan, Marsden, and Murdoch (2006: 195) argue that the local is never pure. Nevertheless, the local and the space of policy-makers encounter one another in the distribution of singularities, in this case the endlessly emergent and contested material manifestations of regional food. These singularities of food territory delineate particular forms of knowledge, organization, politics of affect, and environment. In our view, this illustrates the limits of geographic metaphor and institutions, be they the nation-state or the neoliberal market and its global manifestations. Empirical research places in question the conventions of where and how to situate agrofood studies, spiriting a re-imagination of agriculture and food.

Exploring food territories

We understand the notion of region as a threshold, where assemblages sustain or transform the production, retailing, procurement, and eating of food (e.g., through flavor, taste, and form), the social relations of actors, and the biophysical environment. A region is a zone of sensitivity and sensibility, where shared experience and sensations may operate among humans and nonhumans to encounter and organize the different potentialities that inhabit a particular food-world. A regional perspective provides access to the vibrancy of practice that is located in biophysical space as well as a conjuncture (i.e., the moment or event), where form and content give rise to a body.

As portrayed in an account of industrial salmon farming in Patagonia (Blanco *et al.*, 2015) and the topologies of Avian flu and other biotic diseases (Hinchliff *et al.*, 2012), regions and provisional bodies are in continuous states of coexistence and transformation. Hence, these territories are highly vibrant, contingent, and unpredictable – lying beyond the limits of the global imagination and the imaginaries of institutional society. Food trajectories do not always obey the regulatory rules and norms of the nation-state or the quality demands of a market. This means that any state (country or market) potentially carries the imminent collapse of global processes and reconfiguration of familiar practice. As such, reproduction of the globalization of food is unlikely to continue, and as a result, multiple worlds lie both within and without citizen's control. Not only does this situation place in doubt a state's ability to regulate the market or, more generally, to control what happens within its borders, but moreover, it places in question a state's capacity to cope with the singularities or multiplicities of relationships and materialities associated with agriculture and food.

For the cases presented here, food is a material practice (e.g., production, retailing and consumption) that influences affective global imaginaries (cognition and political practices), where the social and the semantic are interconnected through bodies politic, political cognition, and political affect. Haynes-Conroy and Martin (2010) use identity-based studies to account for the more diffuse forms of social activism in the twentieth and twenty-first centuries, considering bodies and biology as well as minds and meaning in the mobilization of people and their political participation. This orientation requires moving beyond the consideration of the body as a passive material object upon which externally based influences act. Instead, they view the body as an active agent of political engagement and life. This perspective demands a grounded and empirical account of activism as a means of collapsing dualistic representations of reality. It seeks to overcome the separation of mind and body – inside and outside, internal and external – in order to associate how making sense of the world, discriminating among values, and public performance take shape through daily processes of living and being.

As such, race, gender, and class as well as the notions of tradition and modernity, urban and rural, and lay and scientific are at once situated in multiple embodiments or identifications. This is the case in the tagging of conventional and organic production, smells, flavors, and tastes of domestic and foreign cookery as well as quality in popular and fine cuisine (see Arce, 1993). Through associated activism, people come to feel and make sense of their surroundings. In the process they organize the complexities and chaos of contemporary market penetrations of the state, family, and community (see Fisher, 2012). As Hayes-Conroy and Martin (2010) explain, this is not merely an issue of the cognition of educated and progressive activists or foodies, but it also is the product of a bodily kind of resonance activated through *visceral* processes of experience and identification. Rather than universalizing singular moments of sensation, their empirical work on the Slow Food Movement in Canada and the United States leads them to recognize the multiple ways in which minded bodies develop, move, shift, and work. In so doing, bodies become both aligned and involved in the (re)creation of conventional and alternative food movements. In summary, their visceral perspective of food represents a provocative challenge to earlier food studies that tend to perpetuate a separation of mind and body. In contrast, Hayes-Conroy and Martin call attention to minded bodies engaged in the politics of affect.

Drawing on these traditions, Carolan (2011) emphasizes the need to conceptualize food relationally. Together with the aforementioned authors, Carolan's contributions go beyond Phillips in challenging knowledge, cognition, and perception in the tradition of Kantian humanism. In its place rests the lived experience and the affective, sensorial significance of food, as opposed to understanding food as part of a natural order, even if socially constructed. Thus food becomes a multi-sited multiplicity of organisms and things that is integral to (non)human being and becoming. This raises the need for empirical study of the realms of food objects, subjects, and contingencies as a window into their multiple existential forms, processes, and creativities. Before getting ahead of ourselves, however, we need to introduce our understanding of modernization in food and agriculture and its reflexivity, before returning to our central interest in the contemporary states and territories of food in Latin America.

Modern food and its publics

Although diversely meted out, agricultural modernization in Latin America generally has involved land reform and public policy tied to highly creative and ambitious social innovations (Flora and Flora, 1989) and proliferation of an industrial era of technical transformation (i.e., mechanized tillage, severe genetic manipulation of crop varieties, synthetic fertilizers and pesticides, new methods of food preservation and storage, and the arrival of the supermarket as well as the invention of foodstuffs) and an increase in business enterprises, investment in communication infrastructure and the provision of financial resources – all aimed at isolating the rural producer, before seeking to integrate him or her into an emerging modern agricultural economy built on purchased inputs and intermediated, urban-based commercial markets. This work has led to the establishment of national research and extension systems in countries across the planet, supported by a network of specialized commodity centers. In Latin America, the International Center for the Improvement of Maize and Wheat (CIMMYT) pioneered a "Green Revolution" largely based on this model, followed by the International Potato Center (CIP) based in Peru and the International Center for Tropical Agriculture (CIAT) in Colombia.

While results in terms of increasing production per area, providing cheap food for growing urban centers, and reducing or eliminating hunger and poverty were diverse and mixed, from a social perspective, agricultural modernization was extraordinarily successful in establishing an unprecedented market-based, state-led technocratic ideological project based on: (1) commodification of rural life (i.e., placing a price tag on time, space, natural resources), (2) introduction of currency and eventually financial systems (as a means of mediating human and human–object interactions), (3) the social and geographic distancing of markets (thereby separating processes of production and consumption), and (4) reliance on expert-based knowledge and technology (thereby creating conditions for dependence). For Sherwood and colleagues (2015), these four phases of modernization provided the foundation for extraordinary growth in productivity, followed by severe socio-environmental decline in agriculture and food in Latin America. As Beck (2001: 271) explains, "It is not rule-breaking but the rules themselves which 'normalize' the death of species, rivers, or lakes." He described the self-destructive feature of modern societies as "organized irresponsibility."

Environmental decline, pesticide poisoning, and obesity tied to modern food are political issues, because in the current neoliberal climate, health concerns are simplified as primarily economic matters, demanding re-allocation of scarce, highly demanded government resources. Thus, there are professional and financial incentives to consider the unwanted products of modernization as a personal choice or perhaps as self-inflicted action resulting from a combination of a lack of information and capacity and irresponsible behavior. These simplified frames of reference ignore agriculture and food issues as immediate threats to the public economy, health, and the environment. They are also political issues in that they have contributed to increased citizen's political agency (i.e., agroecology, food sovereignty, economic solidarity movements) to defend and advance certain interests over others. In a democratic society, this dynamic is part of an increasing process of politicization within people's rights and claims to well-being. The resulting policies have important financial implications for private and public health care, civil rights, and business practices.

At present, the very issue of citizenship is being actively debated in many parts of Latin America, including in relation to agriculture and food. This is rooted in both the development of science and technology and evolving expectations and patterns of governance. Technological advance and the ongoing growth of the consumption society are opening up new spaces for citizen action and democracy, while at the same time conflicts with conventional expert-led development and government institutions are generating heated public debate and reflexivity, calling into question the very categories of government, science, and citizen. This is most clearly illustrated by the ongoing debates between civil society and the scientific community and government over specific agricultural legislation intended to govern the use of land, water, biodiversity, and markets as well as agricultural production and conservation.

In addition, special attention is required for human–nonhuman relationships and a need to redefine the public as extending beyond the citizen and the consumer. Environmental degradation, mass pesticide poisoning, and overweight/obesity can be seen as unintended effects of the public modernization policy tied to a particular pathway of agricultural development. As we have seen with growing pest outbreaks, decline in bee populations, loss of agrobiodiversity, and growing climate variability tied to global warming, the consequences of modernization shape nonhuman worlds in ways that feedback on humans and their enterprises (i.e., the arrival of the anthropocene). As such, we do not view these consequences as "externalities," but "internalities" of new forms of modernity. We find that the citizenry as well as policy-makers and health care experts commonly overlook or neglect the particularities of contemporary, unwanted food products and foodstuffs (i.e., the health consequences of ultra-processed foods), which include intricate relationships between human and nonhuman entities.

If not dealt with immediately, the social and policy weight of agriculture and food quality will be borne by future generations. In this regard, food quality is comparable to pollution and global warming. Nevertheless, modern food, as part of an increasing politicization at the grassroots level, offers a critical window into the uncomfortable realities of daily life and raises awareness of the dichotomous nature of progress and modernities. This is manifested, for example, in the growth and influence of food movements in Latin America organized for the purpose of uprooting established regimes of practice as well as claiming access to heritage, healthy, equitable and sustainable living, and demonstrating a viable alternative to the dominant discourse that there is only one acceptable modernity in the region (Arce and Long, 2000). In this manner, the political dimensions of agriculture and food are central to rethink food policy. They tend to be associated with the potentials generated by reflexive modernities, in which technocratic and expert-based "systems" appear to be collapsing into the complex pragmatic of daily living and being.

In light of this analysis, we find the need for new thinking in citizenry and public administration that today continues to center on how to bring the public into line with techno-administrative decision-making, thereby neglecting or denying the roles that people already play in the enactment of agriculture and food policy. Here, we ask whether the very nature of the public needs to be redefined to include both human and nonhuman entities. In order to overcome the harmful consequences of modernization, emphasis needs to be placed where it belongs: strengthening new forms of democracy where people's rights, roles and obligations, imaginations, and capacities indeed steer the course of change of public institutions to multiple shared visions of the future, but also human–nonhuman relationships. This view of citizens as authors repositions people at the center of their new experience of what is a family, neighborhood and community today, where they are at once objects of government, but also central subjective agents of government and governance. Addressing the modern multiplicities in Latin America ultimately requires attention to a form of democracy that makes transparent and meaningful the very purposes of citizenry, public space, and government.

Nation-state, market and techno-governance

In this book we view the social sciences in the development of agriculture and food as a field of collusion, collision and competition among the entities dedicated to the fostering of narratives. Currently, both the voices that support and criticize capitalism and neoliberalism have come to naturalize capital, the free market, and the state, understanding each as a driving force of a single vision of necessary modernity.

As Bruce Braun (2015: 1) summarizes, "At a moment when capital presents itself as coexistent with social, political and ecological life, we are reminded of a crucial point: capital is not the source of life, but parasitic on it." We agree with Braun that the conceptual work carried out by capitalist thinkers and their imagination has a radical significance that goes beyond just economics. Capitalism has come to shape imaginations and policy and inform how people think about and experience life, the social realm, and existence. In other words, capital is an entity that is so important that it surpasses ideology, even of its own grave diggers.

Similarly, there is another vision that is also relevant in social sciences literature that highlights the importance of "resistance" and displacement of capital. The advocates of this position have fallen, however, into the trap of trying to provide solutions to the problems of neoliberalism through institutional means. These solutions are generally based on a single unquestioned premise of the centrality of capital, technology, and the nation-state to social change, ignoring the fact that said changes are fundamentally contingent and even accidental. For them, the issue is not just the *lack* of policy imperatives, but the way in which actors gain (or not) access to entities of change.

These visions present us with capitalism and neoliberalism as overwhelmingly social forms and hinder reflection on the importance of criticism on the underlying sources of social justice and the undermining of the environment and natural resources. Nowadays, progressive social sciences in the field of agriculture and food in Latin America are dominated by the rationality of political economy in its different variations. These studies generally explain the notion of poverty as a result of the persistent and growing wave of neoliberal capitalism that plagues the region. This literature is generally inspired by the failures of neoliberalism, in particular, the aim of organizing human activity around the privatization of the social, market deregulation, the reduction of public investment, and the prevalence of the free market, leading to the popular belief that these "lacks" cause the growing socio-political marginalization and environmental degradation in Latin America.

Meanwhile Boas and Gans-Morse (2009) explain that there is no clear definition regarding the conceptual meaning of neoliberalism. Some consider it an economic doctrine, but in light of the diverse violent geopolitical events in Latin America, the ideological element also is important to consider. In the second half of the twentieth century, political developments such as the coup d'état against the Guatemalan government (1954), the Cuban Revolution (1959) and the coup in Chile (1973) caused implosions and explosions – in terms of human rights, democracy, and social, cultural and economic inequalities – that generated considerable criticism of the ensuing wave of neoliberalism in Latin America. More recently, in places such as Bolivia, Ecuador, Nicaragua and Venezuela, the region has experienced the replacement of conservative patriarchs with democratically elected popular authoritarian leaders or *caudillos*. In 2016, Brazil experienced the region's first democratic coup, challenging people's commitment to state representation as well as to its very institutions.

Today, voices of dissent challenge ideological influence within nation-state institutions, especially with regard to decision-making on public policy. This raises questions around public–private partnerships, but above all, around the interests of large companies, which undermine the redistribution of resources and the control of competitiveness between national economic entities, the public definition of efficiency, and regulations on capital flows. In this way, political economy, which in part arises from the moral philosophy of the eighteenth century, is also rooted in Latin American Dependence Theory of the 1960s (Arce, forthcoming).

More recently, the field of political economy has been revised and influenced by "the Third Way" of technocracy (Giddens, 1998), which has contributed to the belief that coexistence between the state planner (social democracy) and free market operations (classical liberalism) is possible. Accordingly, capital is not necessarily considered at odds with social purposes; capitalism is understood as a process that, without being perfect, is all that there is. Therefore, it is argued widely that what is needed is the institutionalization, regulation, and modernization of capitalism, for example through new technologies and incentives promoted by the market. In this view, the role of the state in the social sphere is to promote personal responsibility and foster employment rather than dependency on social welfare. Although social inequalities are immanent to capitalism, the state seeks to promote equal opportunity. The economic objective is not to redistribute income but to increase income for everyone.

Presently, states such as Chile, Ecuador, Argentina, and Mexico have organized around this neoliberal project to assure further integration with financial systems through electronic debit and credit in the name of efficiency and transparency. This political and economic project depends on a modernized public–private bureaucracy that is capable of combining monetary policy, taxes, and investments in order to achieve new degrees of economic growth, while generating tax revenues for public investment. The Third Way represents a conceptual and technocratic model that attempts to depoliticize the social in order to coordinate with private interests in service of the growing aspirations of an urban-based collective that is increasingly organized around a public parade of objects. In this model of consumer sovereignty, economic expansion is seen as a socio-political process outsourced to modernized private companies that, when coordinated institutionally with the needs of state welfare and bureaucracy, resolves the contradictions and conflicts between public and private. This naturalization of the political and social, achieved through administrative capacity and ability of elite social democratic reformers that represent the interests of diverse social groups and their managing contradictions, dominates the socio-political objectives of today's cutting edge food policy.

Followers of the Third Way often prioritize the need for a socially oriented market, and they suggest democratic management for enhanced public consultation and participation as the political manner of dealing with the scarcity and degradation of natural resources and social services. This orientation is geared to favor continuous and sustainable economic growth based on science and technology, but also innovative investments, such as renewable energy and the introduction of biotechnology. In other words, they rely on greater degrees of capitalism and technology as entities that finally can solve the problem of poverty and cyclical crises. The establishment of public service management that prioritizes the interests of the nation-state through efficient implementation and citizen-led policies with a "human face" is seen as a fundamental requisite for socio-natural and economic development. The basis for this orientation is founded in the criticism of the "traditional" interventions and decision-making processes and non-participatory deliberation of nation-states during the period of neoliberal modernization. Additionally, they protest against the policy of public multinationals, such as the agencies of the United Nations, the International Monetary Fund, and the International Labor Organization, as they attempt to establish or correct them.

Ultimately, the Third Way is a state-nationalist stance. The topics that dominate this orientation emphasize the centrality of the nation-state in social life, the importance of regulating economic sectors, controlling monopolies, protecting markets, promoting transparency, citizen participation, and controlling corruption in the public sector. The goal is to facilitate social distribution of wealth and civic responsibilities beyond the benefits for the elite. Latin American social scientists commonly own this view of political economy, especially in reference to the notion of dominant global economic, productive, and technological regimes, such as the global financial reality, the industrial food complex, and the green technology revolution. As per this thinking, different assemblages influence and are influenced by the organization of capital, resources, space, and access, but also by the organization of the workforce.

Often the concepts used in Marxist critiques of capital, capitalism, and neoliberalism are based on academic neo-Marxist movements, which have concentrated their social relevance in fields of study related to development and underdevelopment. These analyses often reduce human experience to a class conflict and the effect of adaptation to social change. The outcome of these conflicts of interest and adaptation to the differences between social classes has led to highlighting the processes of resistance of subordinate social groups, such as farmers, that according to the thesis of the Third Way, mobilize against the socio-political sectors of the ruling regime or system. In the populist version, farmers, and more recently ethnic groups, are understood as historical subjects that have not been absorbed by capitalism, thus representing the existence of the experience of social classes that fight for liberation and the strengthening of different worlds.

Faced by this situation, the implicit position of the Third Way is to establish the existence of a coherent, cohesive society. The answer rests in improving what adherents view as natural functions and roles of citizen groups, dissolving hierarchies and lessening the differences between social classes and developing forms of cooperation and competition with aggressive economies (particularly the United States, Europe and China) and international corporations (e.g. Coca-Cola, McDonald's, Monsanto, and pharmaceutical, mining and petroleum companies) that attempt to achieve a productive homogenization of consumption and regulation across the planet.

Following a half-century of public attention geared at reforming the agricultural sector with technical assistance and modernization of production, food, and markets, it would be a mistake to forget that today's agriculture and food crises have continued and deepened. Hence, we are concerned not only with the problems of the past (e.g., how to feed the public and how to integrate the traditionally marginalized sectors with the benefits of modern society), but also with pressing issues of the present. In particular, champions of the Third Way must reflect on how to organize social and environmental responses that are both scientifically informed and publicly sponsored.

More recently, the focus of political ecology has retained the notion of the primacy of the market and its social orientation, while conserving the importance of a strong state, attempting to politicize contemporary environmental concerns, socio-climatic events, the extraction of resources and the importance of promoting the distribution of power, especially with regard to access and conservation of natural resources. Much of this orientation informs the thinking in "alternative food movements" (Brenni, 2015). The concern lies in the main political or sustainable coexistence between human and biological/ecological configurations. In particular, Brenni studies the possibility of changing the incentives of the market economy. He is especially interested in replacing the notion of maximizing financial return with new values that arise from the importance of environmental and socio-cultural services, and even recovering the spiritual and symbolic beliefs, to thereby establish different forms of political action and production and consumption practices.

In short, political ecology prioritizes socio-environmental issues, such as the need to address social exclusion and anthropocentric degradation and mediate in environmental conflicts. It emphasizes the control and conservation of nature, "environmental rights," and social movements as fundamental aspects of a contemporary socio-political agenda. In this book food brings to bear a sociological understanding on questions of management and access to natural resources, the implementation of agroecology, and the defense of food sovereignty.

A critical reflection leads us to conclude that many of the solutions implemented so far, in terms of public policy, have not resulted in the promised changes. Paradoxically, it appears that public policies have played a central role in generating new problems that are often worse than the original problems that they were meant to solve. Well-known examples in the region include: large scale degradation of soil and water, the erosion and loss of genetic resources, the proliferation of pests, the loss of competitiveness of local and regional food markets, plus some previously unimaginable consequences, such as poisoning and diseases of entire populations due to the effects of pesticides, global warming, and obesity. Our studies show that we are witnessing the development of a new agrofood crisis (Sherwood *et al.*, 2013). In other words, the socio-environmental crises that publics are experiencing in Latin America are not caused by government or scientific neglect or deficient public policy. On the contrary, they commonly are the direct result of the "success" of modernization in agriculture and food.

Empirical studies of contemporary food, agriculture, and social change in Latin America

Throughout Latin America, agroecology, food sovereignty and responsible consumption movements have demanded citizen rights from the state, while at the same time demanding degrees of autonomy from it. In essentially all cases they have lost miserably, as proposals have been rendered technical and hopeless, bureaucratized by actors in the state. For our colleagues in Latin America as elsewhere, in this book we ask: how can a conceptual debate be opened up based on the contribution of social sciences to the empirical realities of food and agriculture?

Building on Csordas' (1994) seminal work on embodiment and experience, the book's questioning of the nation-state and the market is based on an understanding of development in food and agriculture as fundamentally about being and becoming. The territory of what has become a highly fluid state is not limited to geography or institutional structure, but rather to the dynamic situations and boundaries of objects, subjects, and multiplicities in continual interaction (Umans, 2016; Umans and Arce, 2014).

Not only is this territory an *objectivization* of biophysical and social singularity, but it also is a inter-*subjectivization* of space determined and acted upon through a plurality of common sensitivities and sensibilities. We understand this region or space of agrofood encounter and development as a singular-plurality or *multiplicity* composed of lifeworlds that are simultaneously local–global, material–liquid, transient, made up of endlessly nuanced and fragmented boundaries in different states in (dis)solution or being and becoming. As per the material presented here, multiplicities can be found in crops as both a source of food and craft and cultural identity and pesticides as both a solution to pest problems and a source of death or GMOs as a means of controlling nature as well as the creation of new, uncontrollable natures.

In this book, 27 authors working on 14 case studies in seven Latin American countries aim to capture and explore the situated complexity of politics and practices of food (see specific locations in the map, Figure 0.1). This experience is presented through the focal point of diverse forms of modern participation and social activism in civic and political life – all witnessed in real time during different phases of hope and aspiration – to transform and displace agricultural modernization and modern food. While the different contributions address a diversity of themes and conceptual material that defy simple categorization, we have roughly organized the resulting 14 chapters by what we see as largely unexplored regions of critical food studies: food embodiments, corporealities, and assemblages and multiplicities.

The book begins with a series of chapters that present different modern food embodiments. Eleanor Fisher and colleagues study a Chili Cook-Off in Ajijic, Mexico to reveal how organoleptic qualities can give expression to people's changing relations with food, thereby generating new "food lifeways" – namely, ways of being in the world that focus on the interactions between people and things through the medium of food. Their research explores how bodies and material things align with people's sense of dwelling in a locality and their commitment to charitable aims through a public event. A subsequent chapter involves Joan Gross and colleagues' work in constructing a "history of the present" of post-modernization dietary transition in Carchi, Ecuador. This study explores how new materialities, such as ultra-processed food and agrochemicals (as well as roads and vehicles), shape relationships between human bodies and the substances and activities that biologically create them. Camilo Torres and Gerard Verschoor present a provocative, if worrisome case of encounter between traditional Andoque communities in the Colombian Amazon and the mining industry, leading to struggles – physical, ontological and otherwise – over living in abundance and the pursuit of accumulation through the utilization of gold as a precious metal and commodity. This work is followed by Andrew Ofstehage's deep empirical study of North American family farmers that migrated to Brazil's vast *Cerrado* region to become transnational soya producers to reveal the agency, plans, and improvisations that give life to such transformations.

A subsequent collection of chapters begins to address questions of corporeality, involving greater attention to the body and the material world as distributed across materialities, where agencies emerge through practice that acts on daily intentions as well as contingency. Horacio Narváez's research in northern Ecuador explores the daily battle and resulting creativity in bringing to bear the object-subject of water on agriculture and food, leading to the transformation of "a desert into an

FIGURE 0.1 Case study locations

oasis." His work identifies how the preoccupation of present-day development projects with the materiality of irrigation effectively blinds technicians to people's affectivity and underlying socio-biological-material relationships, thereby limiting the potential of development interventions. Meanwhile, Nicolas Vasconcellos shares his work on the tropical region of Santo Domingo, Ecuador on how the arrival of state-supported milking technology is intended to improve the efficiency and hygiene of milk production. His study leads to an exploration of how technological intervention, where animals, machines, peasant-farmers, and experts encounter one another, ultimately leads to unexpected intersubjective relationships of affect, threatening to turn the tables on the predominance of expert knowledge in rural development. Oscar Reyna presents a multi-species study of the Wirikuta Indians in northern Mexico and the Peyote cactus. His work exposes how a cactus' genetic properties have triggered affections and invited humans to become interested in preserving the desert from extraction, leading to an engagement in new forms of biopower, as expressed in nomadism and affirmative politics. Together, this work uniquely explores connections between human and nonhuman realms and calls attention to the intersubjectivity of agriculture and food as well as changing notions of space, territory, and time. Meanwhile, Christine Franke and colleagues arrive on the Island of San Cristóbal in the Galapagos to examine the social construction of paradise vis-à-vis people's daily agriculture and food activity. Despite the common criticism of colonists, their analytical attention to the corporeality of contemporary practice sheds light on the potentiality of existing means and processes of farming and marketing by colonists as a hidden resource for addressing pressing livelihood and conservation interests, ultimately leading to a questioning of the notion of "green citizenship" and a call for an ontological shift towards "corporeal citizenship."

A final series of chapters explores the unfolding agrofood assemblages and multiplicities tied to state–civil society institutional encounters over agricultural production, artisanship, and markets as well as the social viscerality of food sensations. Flavia Charao-Márques and colleagues' long-term study examines the unfolding agencies involved in giving rise to agroecology networks in Brazil. Their contribution investigates how critical and practical action against dominant modes of agriculture and food consumption has fostered creative associations between practices, agencies, and materialities, giving birth to creative forms of activism and association that over time have become highly influential in championing agroecology vis-à-vis science-, state-, and industry-led development in Brazil. Their work highlights the multiplicity of materials and socials involved in social change and development as well as the endless potential for creative, transformative relational flows in agriculture and food. Meanwhile, Laura Viteri presents the affective interactions between humans and other materialities (e.g., milk, machines, factories and centers for the retired) during a public procurement experience that led to an employee takeover of the New Dawn Milk Cooperative in Mar del Plata, Argentina. By means of ethnographic travel through dairy production and delivery to retirement community centers, the research explores how a public procurement contract emerged from the cooperative's private need to sell and a public official's need to demonstrate equitable and sovereign policy, leading to a reflection on the importance of affects and intersubjectivities in transforming collaborative associations.

In their study of wheat in Bio-Bio, Chile, Paola Silva and colleagues explore wheat as a socio-material "public" assemblage that reveals how things acquire agency in a territory. Beyond conventional interests of wheat breeding programs that primarily focus on production for flour, their research finds families involved in the maintenance of traditional varieties, leading them to highlight the neglected, "imperceptible" outputs of wheat that are often uncertain due to the randomness of the forces of nature (i.e., earthquakes) and site (straw handicrafts) that are central to territorial development as well as potential for change. The authors find that grounded people–wheat associations and specificities (practices), multiplicities (alliances between humans and nonhumans) and resulting assemblages fundamentally involve relationships and relationship-making, such as those involved in handicraft and food. They argue that such associations, assemblages, and relationships are central to the actualities and potentialities of rural life and livelihood.

Charlynne Curiel in Oaxaca, Mexico and Gustavo Blanco and colleagues in Valdivia, Chile present respective studies based on their experience as participants in urban-based efforts organized around ideals of responsible consumption. Curiel's work is on the transformative experience of taking part in a joint food production-purchasing group, the *Tenate Básico*. Inspired by the principles of bio-intensive farming, the study follows a series of encounters to reveal the effects and affects of the participant's intensified food activity, leading to new relational forms among the social, the natural, and the material and ensuing potentialities. Meanwhile, Blanco *et al.* highlight the centrality of affects in the process of creating a new subject (the responsible consumer) and new objects (the food cooperative and fair food), leading to a call for moving beyond structural and materialist understanding of an assemblage. The authors find that La Manzana, as a food assemblage, generates an intrinsic value of intensifying existence that thrives on creative and novel associations between heterogeneous elements that open new ways of imagining, producing, and experiencing food-related worlds.

While Sherwood and colleagues also engage responsible consumption, they take a different tack: the "sensation" of food movements. Through reflective study as researchers–food activists, they document the activity of Ecuador's Agroecology Collective (*Colectivo*) and the Movement for Economic and Social Solidarity (MESSE). In so doing, they adopt the view that food is part of an endless process of socio-material embodiment that is best explained through people's feelings, practices, and performances. In particular, their research focuses on the joint Colectivo-MESSE "250 Thousand Families" campaign – an ambitious effort to recruit a critical mass to invest in responsible consumption as a vehicle for agroecology. Through ethnography and reflective practice as participants in a series of sensorial workshops, they find that flavor becomes a transformative moment and taste a transformative space, working in solution and precipitating in practice seeming binaries of conception/action, theory/practice, mind/body.

Taken together, the experiences invite a re-thinking of the notion of the public as well as agriculture and food as an object and commodity. In the conclusion, we draw on the insights of the chapters to provide a remedy to what we see as a pressing theoretical, normative, and methodological agenda for a more empirical and critical food studies.

References

Arce, A. (1993). New tastes in industrialized countries and transformations in the Latin-American countryside: an introduction to the local cases of Mexico and Chile. *International Journal of Sociology and Food*, 3, 48–70.

Arce, A. (2009). Sources and expressions of power in global food coordination and rural sites: domination, counter-domination and alternatives. *International Journal of Sociology of Agriculture and Food*, 16(2), 2–20.

Arce, A. (forthcoming). Dependency theory and underdevelopment. In Callon, H. *The International Encyclopaedia of Anthropology*. New York: Wiley Blackwell..

Arce, A and Long, N. (eds). (2000). *Anthropology, Development and Modernities. Exploring Discourses, Counter-Tendencies and Violence*. London and New York: Routledge.

Arce, A. and Marsden, T. (1993). The social construction of food: a research agenda. *Economic Geography*, 69, 293–312.

Beck, U. (2001). Ecological questions in a framework of manufactured uncertainties. In S. Seidman and J.C. Alexander (eds), *The New Social Theory Reader*. London: Routledge, 267–275.

Blanco, G., Arce, A. and Fisher, E. (2015). Becoming a region, becoming global, becoming imperceptible: territorialising salmon in Chilean Patagonia. *Journal of Rural Studies*, 42, 179–190.

Boas, T.C. and Gans-Morse, J. (2009). Neoliberalism: from new liberal philosophy to anti-liberal slogan. *Studies in Comparative International Development*, 44(2), 137–161.

Bonanno, A. (1993). Agro-food sector and the transnational state: the case of the EC. *Political Geography*, 12(4), 341–360.

Bonanno, A. and Constance, D.H. (1996). *Caught in the Net: The Global Tuna Industry, Environmentalism and the State*. Lawrence, KS: University Press of Kansas.

Braun, B. (2015). New materialisms and neoliberal natures. *Antipode*, 47(1), 1–14.

Brenni, C. (2015). Where are the local communities? Food sovereignty discourse on the international agrobiodiversity conservation strategies. In A. Trauger (ed.), *Food Sovereignty in International Context: Discourse, Politics and Practice of Place*. London: Earthscan, 15–34.

Carolan, M. (2011). *Embodied Food Politics*. Burlington, VT: Ashgate Publishing.

Csordas, T. (ed.) (1994). *Embodiments and Experience: The Existential Ground of Culture and Self*. Cambridge: Cambridge University Press.

DuPuis, E.M. (2002). *Nature's Perfect Food: How Milk Became America's Drink*. New York and London: New York University Press.

Fisher, E. (2012). "The fair trade nation": market-oriented development in developed European regions. *Human Organization*, 71(3), 225–267.

Fisher, E. and Sheppard, H. (2012). Pushing the boundaries of the social: private agri-food standards and the governance of fair trade in European public procurement. *International Journal of Sociology and Food*, 20(1), 31–49.

Flora, C.B. and Flora, J.L. (1989). An historical perspective on institutional transfer. In L. Compton (eds), *The Transformation of International Agricultural Research and Development*. Boulder, CO and London: Lynne Rienner Publishers, 7–32.

Fonte, M. (2010). Naming food after places: food relocalisation and knowledge dynamics in rural development. In F. Maria and A. Papadopoulos (eds), *Perspectives on Rural Policy and Planning*. London: Ashgate, 1–35.

Friedland, W., Busch, L., Buttel, F.H. and Rudy, A.P. (eds). (1991). *Towards a New Political Economy of Agriculture*. Boulder, CO: Westview Press.

Friedmann, H. (1993). The political economy of food: a global crisis. *New Left Review*, 197, 29–57.

Friedmann, H. and McMichael, P. (1989). Agriculture and the state system: the rise and decline of national agricultures: 1870 to the present. *Sociología Ruralis*, 29, 93–117.

Giddens, A. (1998). *The Third Way: The Renewal of Social Democracy*. London: Polity Press.

Goodman, D. and Redclift, M. (1981). *From Peasant to Proletarian: Capitalist Development and Agrarian Transitions*. Oxford: Blackwell.

Goodman, D., Sorj, B. and Wilkinson, J. (1987). *From Farming to Biotechnology: A Theory of Agro-industrial Development*. Oxford: Blackwell.

Gronow, J. (1997). *The Sociology of Taste*. London and New York: Routledge.

Haynes-Conroy, A. and Martin, D.G. (2010). Mobilising bodies: visceral identification in the slow food movement. *Transaction of the Institute of British Geographers*, 35(2), 269–281.

Hinchliff, S., Allen, J., Lavau, S., Bingham, N. and Carter, S. (2012). Biosecurity and the topologies of infected life: from borderlines to borderlands. *Transactions of the Institute of British Geographers*, 38(4), 531–543.

Lash, S. and Urry, J. (1994). *Economies of signs and space*. London: Sage Publications.

Lowe. P., Marsden, T., and Whatmore, S. (eds). (1994). *Regulating Agriculture*. Critical Perspectives on Rural Change Series. London: David Fulton.

Marsden, T. and Little, J. (eds). (1990). *Political, Social and Economic Perspectives on the International Food System*. Aldershot: Avebury.

Massumi, B. (1996). Becoming-deleuzian. *Environment and Planning D: Society and Space*, 14, 395–406.

McMichael, P.D. (ed.) (1994). *The Global Restructuring of Agro-Food Systems*. Ithaca, NY: Cornell University Press.

Mol, A.P.J. (2001). *Globalization and Environmental Reform: The Ecological Modernization of the Global Economy*. Cambridge, MA: MIT Press.

Morgan, K., Marsden, T., and Murdoch, J. (2006). *Worlds of Food: Place, Power, and Provenance in the Food Chain*. Oxford: Oxford University Press.

Phillips, L. (2006). Food and globalization, *Annual Review of Anthropology*, 35, 37–57.

Ploeg, J.D. van der. (2003). *The Virtual Farmer: Past, Present, and Future of the Dutch Peasantry*. Assen, the Netherlands: Konenklijke van Gorcum.

Sanderson, S. (1985). The new internationalization of agriculture in the Americas. In S. Sanderson (ed.), *The Americas in the New International Division of Labor*. New York: Holmes and Meir, 460–448.

Sherwood, S., Arce, A., Berti, P., Borja, R., Oyarzun, P. and Bekkering, E. (2013). Tackling the new materialities: modern food and counter-movements in Ecuador. *Food Policy*, 41, 1–10.

Sherwood, S., Paredes, M., Gross, J. and Hammer, M. (2015). The future of sustainability as a product of the present: lessons from modern food in Ecuador. *Rivisita de Studi sulla Sostenibilita*, 21, 83–103.

Tsing, A.L. (2015). *The Mushroom at the End of the World: On the Possibility of Life in Capitalist Ruins*. Princeton, NJ: Princeton University Press.

Umans, L. (2016). *On the Edge of Fluidity: International Cooperation in Turbulent Times.* PhD Thesis. Wageningen University.

Umans, L. and Arce, A. (2014). Fixing rural development cooperation? Not in situations involving blurring and fluidity. *Journal of Rural Studies*, 34, 337–344.

Viteri, L.M. and Arce, A. (2013). The negotiation of quality standards: a social interactionist approach to fruit and vegetable distribution in Argentina. *International Journal of Sociology and Food*, 20(1), 127–146.

1

EMBODIMENT AND REFLEXIVITY

Gaining insight into food lifeways through the chili cook-off in Ajijic, Mexico

Eleanor Fisher, Alberto Arce, and F. Vladimir Díaz Copado

Introduction

An important issue for food studies in Latin America is how to describe and analyze the ways in which food becomes embodied within lived experience. The introduction and chapters within this volume highlight how a range of food predicaments have emerged across Latin America, as new dietary practices and organoleptic encounters give expression to people's transforming relationships with food. We would argue that processes of embodiment are at the heart of this change, encompassing people's corporeal engagement with what they eat and how food stimulates reflection on the relationships people hold to one another, to places and to material things. Phrased another way, we see these processes of embodiment as part of emerging food lifeways that reveal the changing dynamics of how food is produced, collected, exchanged, distributed, desired, eaten, thrown away, and debated over.

Focusing on the issue of embodiment and its implications for contemporary food issues provides food for thought to enrich debates on the materialization of food politics (Carolan, 2011). Moreover, empirically, a focus on embodiment can act as an entry point for documenting practices, meanings, and actions that exist around food as a form of active materiality that mobilizes bodies, minds, relationships, and objects. Following this line of thinking, in this chapter we focus on the embodiment of food within lived experience by taking the case of the Mexican National Chili Cook-off competition in Ajijic, Jalisco State in Mexico, an event that incorporates bodily judgments about food and in so doing reveals wider dynamics about civic action and the public domain. Through this example we demonstrate how a food-affect-resonance becomes apparent within "food lifeways," a concept elaborated below, and gives expression to ways of being in the world that focus on the interactions between people and things through the medium of food. Here bodies and material things align with people's sense of dwelling in a locality and their commitment to charitable aims through a public event.

On food, embodiment, and assemblages

The notion of embodiment has anthropological antecedents in the actor-oriented approach developed by the Wageningen group under Norman Long in the 1980s and 1990s, whereby the conceptual treatment of ethnographic data depicted multiple ontological realities (Arce and Long, 2000; Long and Long, 1992). More than two decades later, the actor-oriented approach continues to have relevance for the anthropology of food in terms of the recognition of the significance of ontology for understanding the role of food in everyday life. Nevertheless, a critical view that "decenters the question of the subject onto the question of subjectivity" (Guattari, 2006: 22) has shifted thinking on food and food movements in development, as "inducer-producers of salient, public effects" (Bennett, 2010: 39). While an actor-oriented approach emphasizes the social through focusing on human actors, new materialist thinking has drawn the nonhuman dimension to the fore, raising questions about what this means for the social (e.g., De Landa, 2006; Ingold, 2011; Krarup and Blok, 2011).

New materialist thinking has challenged the use of dualisms such as subject and object, nature and culture, and public and private. In this respect the work of Deleuze and Guattari (1987), among others, has been used to demonstrate how human–nonhuman entities become part of a taken-for-granted world (Braidotti, 2013; De Landa, 2006). This world is portrayed as immanent and non-linear in a way that reworks and "breaks through" dualisms (Van der Tuin and Dolphijn, 2010). With our colleague, Gustavo Blanco, we have used assemblage theory to explore the immanent becoming of a salmon-producing region in Chilean Patagonia (Blanco, Arce, and Fisher, 2015). A similar challenge to dualist thinking is also found within actor network theory, where emphasis is placed on the co-constitution of nature and culture (Müller and Schurr, 2016).

While different, in both new materialist thinking and actor network theory, we see expressions of what Braidotti (2015) characterizes as change in the paradigm in which we are living whereby the concept of the human emerges as a question. Focusing on the notion of embodiment, and what this means in an age where the very fabric of food, and what it is to be animal, vegetable or human, cannot be taken for granted, the work of Csordas is relevant. Csordas (1990, 1994) saw the collapse of dualities as framing the central methodological issue of embodiment, permitting investigation of how cultural objects (including ourselves) are constituted as objectified through the indeterminacy and flux of cultural life. Thus, lived experience and culture are the focus of bodily being in the world (Csordas, 1999).

Embodiment becomes not about "the body" as such, but the "phenomenon of the body" and the part it plays within the intersubjective ground of lived experience. The body therefore amalgamates the cultural and the material. In this line, and in order to rethink issues of agency and citizenship, Gabrielson and Parody (2010: 380) use the metaphor of the "leaky body" to envisage bodies as "porous, plural and connected" to social and natural worlds. Through the notion of corporeal citizenship they suggest that a focus on the body can offer a means to rethink ideas about agency and citizenship by necessitating recognition of the common materiality of the human being and the natural world, acknowledging the places bodies inhabit and diversity of human–nonhuman relations (see also Hayes-Conroy and Martin, 2010). For this chapter, reflecting on agency and citizenship is important because they emerge from the lived experience that gives rise to new food lifeways in Latin American countries that are realized within or against existing social orderings and public legacies. These public legacies are incorporated into assemblages that configure practices, forms, and bodies in ways that confer agency upon entities within a territory (see Arce, 2010; Blanco, Arce, and Fisher, 2015; Dewey, 1954).

Our use of the term food lifeways is rooted in the thinking of Deleuze and Guattari (1987: 232–309) and its influence on the work of Ingold (2011) and Tsing (2015), who both link processes of becoming to expressions of being in the world through the life course. Thus, Ingold argues (ibid: 69), beings (human and nonhuman) are relationally constituted, building a trail along which life is lived, always in the throes of becoming without reaching a destination. While for Tsing (ibid: 23), ways of being are the emergent effects of human–nonhuman encounters: "assemblages don't just gather lifeways; they make them."

For us, the concept of food lifeways places emphasis on the embodiment of food within lived experience, as it gives expression to the combination of corporality, affect, and reflexivity, entwining social and natural worlds through practices, knowledge, and technology. In so doing, new relationships and alliances are formed at an interface whereby entities ("real" or not) construct, reconstruct, and dismantle themselves within processes of becoming that constitute a sociality and corporeality different from neoliberal individualism because it is constituted by the fact of actors being in the world. Following Deleuze and Guattari (1987), we refer to this process as inter-subjectivity between entities understood as capable of generating influential degrees of potential, power and force, with the experience of creativity available not just to the human, but also to the nonhuman other. In this vein, we turn to our case,[1] the chili cook-off competition in Ajijic and ask a simple question, namely what does this focus on food reveal about lived experience in Ajijic?

The chili cook-off

The chili cook-off, a contest to cook chili – a stew based on meat, red chili peppers, and spices – that also serves as a charity fundraising event, has origins in the Americas. While rooted in an American/Mexican ethos, annual cook-offs are held around the world and raise many millions of dollars for charity. For example, in the United Kingdom "chili fiestas" and "chili festivals" (incorporating chili cook-offs and hot chili eating contests e.g., "man vs. food") have become popular summer events led by the UK Chilli Cook-off Association and local groups, with annual festivals around the country that combine stereotypical symbols of Mexico (chilis, ponchos, and sombreros), the United States (cowboy hats, denim and sheriff badges), and the United Kingdom (Union Jacks and tea stalls).[2]

Against this background, a long-established competition is the Mexico National Chili Cook-off in Ajijic, which started in 1978.[3] Participation in the Ajijic cook-off provides a focus for different groups of people (retirees from the United States, Canada and Europe as well as Mexicans from Jalisco) to engage publicly together, revealing how the materiality of food crosses social, national, and community boundaries to constitute a leaky corporality that assembles foreign retirees, local inhabitants and a legion of chefs, and avid bodies to test and judge chili concoctions through a competitive public event.

Multiple origins and contested legitimacy

The history of the dish, chili, and the chili cook-off competition is a contested one, with disputes over the origin of chili (Mexican, Texan, or neither) and organizational splits in the international cook-off movement giving rise to different claims over legitimacy. It is reputed that the first known chili cook-off was held in 1952 in Dallas, Texas. However a chili cook-off in 1967 in Terlinga, Texas is considered the beginning of a series of annual contests (Ritchey, 2007).

Organizers of the cook-off in Texas later split up to form their own organizations, which remain internationally influential. In 1970, the International Chili Society (ICS) was formed to carry out the competition in California. A split in 1974 produced the Chili Appreciation Society (CAS), now CAS International (CASI); while in 1983, the Original Terlinga International Frank X. Tolbert-Wick Fowler Memorial Championship Chili Cook-off, Inc. (TOLBERT) was formed. Today each organization has many members and carries out its own annual championships, with separate rules for how the events are conducted by member groups at events in different places.

To confuse matters further, other cook-offs take place outside these organizations. For instance, the annual men's "Chilympiad" in San Marcos, Texas, and the "Hell Hath No Fury like a Woman Scorned Cook-off" in response to the Chilympiad's male exclusivity. The final example concerns our case, the chili cook-off of Ajijic, which originally started as a member of the ICS but became independent in 2009. All these claims of origin emerging from organizational alliances and splits – with perhaps no single origin for chili or the cook-off – and contests over legitimacy reveal how an organizational entity acquires a dense social and mobile materiality of significance to "chili practitioners" that becomes embodied through public engagement at events repeated at different sites and across territorial boundaries, building particular food lifeways for those concerned.

The Ajijic Chili Cook-off

Ajijic is a town on Lake Chapala whose scenic location has long attracted foreigners to settle in the area. In the 1990s, the size of this foreign population grew significantly, stimulated by changes in law that lifted restrictions on foreign investment in land and property and by trade agreements that facilitated the importing of homeland products favored by retirees. Against this background, in 1978, the chili cook-off was started by a small group of foreign settlers guided by an American woman whose motivation for starting the cook-off was to raise funds for local Ajijic charities. In the beginning, one of the founders, Anita, traveled to the United States to find out about running a chili cook-off. Subsequent support from the ICS and foreign cooks was vital because people in the town did not know how to hold a cook-off.

Since 1978, the cook-off has grown to be the largest charity fundraising event in the area. It is run by approximately 150 volunteers, who are mainly from the foreign community but include a small group of Mexicans from Chapala Lakeside. Sponsorship from local businesses has contributed to growth, the main starting sponsor being Grupo Modelo, a beer company where the Ajijic cook-off founder once worked, and today sponsoring businesses include Tecate, Coca-Cola, Heineken and eSun Energy.

Approximately eight charities, including those working with needy children, their families, the disabled and elderly, are supported by the cook-off, with each charity attending the annual event to provide information about their services and to sell food to visitors.[4] Also, at the event companies have stands and there are handicraft, art, and clothes vendors. Visitors can look around the exhibits and eat and drink, while enjoying music and dance performances. Performances include rock and roll bands, children doing modern freestyle dancing, traditional *mariachi* singers, the *Ballet Folklórico,* and the opportunity to learn US line dancing led by the Lake Chapala Society. One stand includes animals such as corn snakes, South American raccoons, and falcons that exhibitors can pick up and handle.[5] Hourly charity fundraising raffles are held to raise money for charity and to provide prizes from many local restaurants and services.

When the cook-off started, membership of the ICS meant that in return for an annual fee the organizers were provided with regulations, which included: a list of permissible ingredients for making chili; the opportunity for winning cooks to participate as representatives of Mexico in the international final competition in Reno, Nevada; and finally, that judges could vote for competing dishes in Ajijic and were given the possibility of being invited to judge international cook-offs. The Ajijic cook-off remained an ICS member until 2009, when membership was discontinued. However, it continues to follow similar regulations, and foreign members of the ICS continue to participate in the contest.

The contemporary cook-off in Ajijic has five competition categories for chili and one category for margaritas. These categories are: "Jalisco State Chili Rojo" for cooking "Traditional Red Chili" and "Jalisco State Chili Verde" for cooking green chili – both categories open only to competitors from Jalisco State; "Salsa" for cooking sauce, open to everyone with no restrictions on what to cook; Mexican National Chili, for competitors from any part of Mexico to cook "Traditional Red Chili"; and, the "People's Choice" for cooking chili and salsa by home cooks from Chapala Lakeside and local chefs from Chapala's restaurants. A local judge awards a "People's Choice" category, determined by the public taking part in the event and using a "tasting kit" (a spoon, napkin, container, and ballot paper) that can be bought for 10 pesos (US$ 1).

As the chili cook-off has become established in Ajijic over a period of decades, we witness how people's engagement with food reflects wider processes of change, as the transnational lifeways of foreign retirees have brought the event to Ajijic. In the process, local meanings, practices, traditions and expressions of affect are incorporated into the public event.

The thirtieth anniversary cook-off in Ajijic (2008)

The 2008 cook-off was held in Tobolandia, a waterpark in Ajijic. In addition to the cooking tents, there were about 50 booths for companies and vendors and about 20 booths for charities and non-profit organizations.

Co-author, Díaz Copado, attended the 2008 competition and describes how at the entrance of the "People's Choice" tent there was a ticket booth where a queue of visitors waited to buy their tasting kits. Around the tent were booths of chefs from the locality, with either their name or a restaurant banner, each chef supported by a spouse and/or team. In a different line of tents, the ICS' cooks were competing in the category, "Jalisco State Chili Verde." Among these cooks were some who travel to different countries to participate annually in chili cook-offs; they are known in Ajijic as "International ICS Chili Cook-off" participants. In 2008, they included a couple that had competed in different cities in the United States, including Hawaii. They said the chili cook-off of Ajijic is well known internationally within the cook-off fraternity, due to its "Mexican style," which contrasts with a more typical Texan style. Mariachi singers and the Ballet Folklórico, among other attractions, form this "Mexican style."

Conversations held at the 2008 competition elicited different responses about the origin of chili. For the foreign cooks, both the dish of chili and the cook-off are Texan; one motivation for participation was to perpetuate a Texan tradition. Nevertheless, some Mexican cooks considered chili a Mexican dish. One Mexican cook, Victor, explained that he participates in the event to prove to foreigners and the local public that chili is Mexican. He elaborated:

> I was really irritated that foreigners came to Mexico to tell and judge how to cook with chili, for example, how to cook the carne con chile, salsa, etc. Imagine that! Mexicans have been cooking with chili since the Aztecs and Mayans in pre-Columbian times. … somehow these foreigners arrive and say that what they call Chili is a Texan dish … [they] … make this contest where

> they impose the rules for cooking *chile con carne*, what they call Chili, and where they even tell you the ingredients you must or must not use. Then they judge the dishes; and in the end the winners are always the foreign cooks ... this is changing, there is more Mexican participation; however it is still a contest based on foreign rules, with mainly foreign cooks, and with mainly foreign winners for cooking a Mexican dish. For these reasons the organizers needed to include the category "Mexican Cook" because they were feeling it had been a bit unfair that foreigners instead of Mexicans had been representing Mexico for many years in the world championship in the US.

Victor recognized, however, that another reason for participation in the chili cook-off was to support foreigners in their "wish of making Ajijic a place where foreigners and locals join as one to bring local development through charitable activities." Another man, José, a performer of the *Huentli* ritual in the town,[6] had similar reasons for attending. He saw it as a good opportunity to help the needy and to contribute to the foreign organizers' objective of uniting foreign and local communities and to bring local development by supporting charities.

While a focus on chili suggests the cook-off is a foreign affair, people (both Mexican and foreign) also attend the event because of the traditional music and dancing. Guadalupe, a 54-year-old Mexican woman from the town of Chapala, explained:

> I like to come to the chile cook-off because I like to see the Ballet Folklórico. I like their performance, their dresses, and the music. I always get goose-bumps when I see them dancing, it makes me feel happy, it makes me feel very Mexican and even it encourages me to dance – ha, ha, ha.

George, a 75-year-old man from Michigan, explained:

> You see me as a blond, but I think I am Mexican because I always like to be part of the Mexican celebrations. I come to the Chili Cook-off every year because here you can see all the great things Mexicans do, their traditions, the Mexican men (the Mariachi) wearing big hats and singing. This is the reason why I moved to Ajijic, to live the Mexican life.

Reflecting on these accounts, processes of embodiment amalgamate the cultural within itself, the dish carne con chili is long-established within local cooking traditions while chili con carne has roots in southern American cooking traditions. Taken together with traditional Mexican music and different forms of dancing, all stimulate the re-composition of elements that give rise to affect – taste, bodily sensation, sound, movement – each playing out elements of difference and commonality within new food lifeways in the locality of Ajijic. Divisions and continuities between and within local inhabitants and foreigners are revealed in how a competition, originally patronized exclusively by foreigners, now incorporates a wider group. In this respect, food acts as a vehicle that assembles elements of creativity and corporeality, linked to the materiality of the chili cook-off.

Charitable giving, estate agents, and comuneros

At the chili cook-off people may come together who are members of social groups in Ajijic between which there is normally strife. One example is between realtors (estate agents) and the indigenous community. There are long-standing tensions between the indigenous community of Ajijic (*Comunidad Indígena de Ajijic*), known locally as *comuneros* who have sought to protect the status of communal land threatened by residential development. In the 1990s, activism by comuneros against the municipal project of transforming Ajijic into a foreign retirement destination, as promoted by realtors, stopped the construction of some residential development.

For the 2008 cook-off, realtor and resident of Ajijic, Ana, attended with comunero Alfredo. As they explained, they were together because they wanted to contribute to forging Ajijic as a town where foreigners and locals work together to stimulate local development through charity fundraising events. Alfredo said that he had joined together with Ana to show people that in spite of the differences between the comunidad indígena de Ajijic and some realtors, they can still be together for the sake of the needy of the lakeside. This they could do by buying tickets and attending the event, buying tasting kits and raffle tickets. The perceived value of the cook-off for joining together of different communities is explicitly portrayed in the official chili cook-off T-shirt emblazoned with the US, Mexican, and Canadian flags and the event's motto "Mexican National Chili Cook-off. *Juntos Logramos Todo* (together we achieve everything)".

The focus of "achieving everything" at the cook-off is charitable giving. Alongside the cooking tents are booths promoting charities at the event, including the local charities that will receive funding from the cook-off. The cook-off in Ajijic is considered by organizers, participants, and visitors as a symbol of charity, which can be the main reason individuals participate in the event as well as a company's motivation in sponsoring it. Most of the proceeds collected from the Ajijic cook-off are donated to eight selected charities after the event. For example, in 2016, seven sponsored charities received 32,000 pesos each (approximately US$ 1,760). To this end, each booth at the charity provides information about its services and the needs it is addressing. The eight charities also sold food (hamburgers, hotdogs, spareribs, donuts, cookies, etc.) and one even sold second-hand furniture. The charitable proceeds from the event have grown year by year as the event itself has grown – in an interview, the event's founder estimated that the event raised approximately US$ 5,000 in 1978 and about US$ 45,000 in 2009. After costs, a profit of US$ 28,000 was equally distributed among the eight charities.

Marilú, a representative of the Villa Infantil orphanage in Chapala Lakeside, was volunteering as an exhibitor at the cook-off because of the affection she feels for the orphanage's children and their need for help and support. She described how the chili cook-off in Ajijic gathers many charities together in its grounds enabling charity representatives to talk directly to each other to learn what donated items each charity needs and what they have in surplus to exchange among themselves. The relationships with foreign charities are also important. Marilú explained how some time ago a Christian organization from the United States attended the cook-off and met representatives of another Chapala Lakeside charity, Misión San Pablo. After visiting its facilities, they decided to help and have formed a long-term relationship, periodically donating money and items from the United States.

In the accounts of peoples' support for charity, we can see how food can help build relationships, using people's sense of being in the world to become part of the local public. This process has contradictions, but it is through these connections that people and things become part of the public, with charity becoming a specific kind of civic action that gives meaning to life connections in the world.

Chili, taste, and judgment

The cook-off reveals people's feelings towards food (of attachment, sympathy, and affection), which are bound to their bodily senses (the smell and taste of chili and hearing the mariachis). Of these senses, that of taste yields significant influence over the social dynamics around a central activity of the cook-off, namely the judging of competing chili dishes. According to interviewees, the judges' taste is such an important issue not only because it identifies the winners but because how they vote can influence the number of competing cooks and visitors attending the event, with consequent impact on the level of charitable donations that are raised. In particular, the judges' taste affects the number of Mexican cooks and visitors that come to the cook-off because how they vote can reveal preferences for flavors that people are most familiar with in their everyday lives.

According to the Ajijic National Chili Cook-off website, "Chili is defined as any kind of meat or combination of meats cooked with chili peppers, various other spices and other ingredients. Beans are acceptable." Major considerations for judging the chili are: good chili flavor, texture of the meat, consistency, blend of spices, aroma and color. So judges at the cook-off, some of whom have long experience evaluating dishes at international cook-off events, make their judgment based on these criteria. During the tasting itself they cannot deliberately favor a particular competitor because the dishes are numbered anonymously, creating the impression that it is only the judges' taste that matters and determines who wins. This taste, however, is not "objective"; local Mexican participants allege that they lose the competition because Mexican cooks' dishes are influenced by the Mexican "carne con chile," whereas the ICS judges in Ajijic are mostly foreigners who demonstrate a bias for the US "chili con carne" (for example, in 2008, 11 out of 13 judges were non-Mexican).

Over time, allegations of bias demotivated local Mexican inhabitants of Ajijic to attend the competition. However, this changed when the "People's Choice" category of dish was introduced because Mexican visitors to the event could vote, using their tasting kits, and tend to favor dishes that taste more like "carne con chile." The move to incorporate Mexican tasters from the public brought more Mexican competitors and visitors, which led to higher charitable donations. In this example, a bodily sense integral to the pleasure of food stimulates people's agency in terms of how they engage with the cook-off competition.

Beyond the sense of taste, people's engagement with their bodily sensations and feelings of affection for the chili cook-off's symbols are significant: people's feelings of nationalism for chili and chili, feelings of sympathy and affection for the needy, the feeling of enjoyment gained by listening to mariachis or by seeing the performance of the Ballet Folklórico, the enjoyment foreigners gain from the "Mexican touch" or local inhabitants from being reminded of the "Mexican life" have each been central to bringing organizers, participants, and visitors together at the event. This includes feelings of affect, which encourage many people to volunteer to raise funds for local charities. As Anita said, "The Chili Cook-off would not exist without the good feelings that motivate people to volunteer for the chili cook-off … This event is organized and carried out by volunteers." This volunteering has been the galvanizing force that has kept the event running for 38 years (to date in 2016) generating the capacity to transform Ajijic, and, in particular, the life of people dependent on local charities.

The way food stimulates senses that embody people's engagement with the chili cook-off draws the public to the event and shapes the practices and relations between people and between people and things during the course of the competition. In this respect, we witness the capacity to affect and be affected by the senses, amalgamating the cultural and the material, with agency granted to nonhuman entities. This can be seen for example through the way chili "brought" someone to the event, or through the culinary traditions in which the flavors of competing dishes has an effect on the judges' bodies through their sense of taste, and makes them declare one dish or another the winner. Such bodily food-affect-resonances have been essential for the growth of the competition and its contribution to Ajijic through millions of pesos donated to Chapala Lakeside charities.

Discussion and conclusions

In this chapter we have to consider how food is embodied within lived experience in ways that interconnect affect and corporality, the social and the material, giving expression to the public through lives shared around food. The example of the chili cook-off competition in Ajijic, Mexico, was used to demonstrate how a food lifeway emerges in a way that is both specific and continuously evolving, as foreign retirees and different groups of local inhabitants build lives and relationships within the locality of Ajijic. The chili cook-off gives expression to the relationships people hold to one another and to nonhuman elements of which, most obviously, chili peppers and the dish chili con carne/carne con chili are central players. Bodily being in the world is integral to this process, as expressed through organoleptic experience as different tastes for chili emerge from a leaky corporality that exposes wider social relational fault-lines within Ajijic and across transnational boundaries.

A focus on the chili cook-off has served to show how food becomes embodied in a public event that acts as an intersection between food, people, and things, generating sociability and ongoing public action (charitable giving). Within the food lifeway, the cook-off acts as a carrier of affect about culinary traditions that stimulates bodily agency, emerging as something that is material and sensorial and always present in the making of the event. Such agency, and the inter-subjectivity it implies, is partial and exists in real time through the unfolding materiality of bodies within a food lifeway. By implication, the embodiment of food is not a simple extension and accumulation of actors' experience, of discourses or a determination of the past over the present. It is the bodily being in the world that facilitates awareness of what we are as quotidian entities, reminding us that we are not fully aware of how we become what we are. Beans, chilies, taste, touch, sound and charitable sentiment become woven into entities that form an embodied sociality giving expression to public food practices.

The chili cook-off competition provides one opportunity to think about food and agency as partial forces. The vibrancy of life makes it an active combination in the process of embodiment of social, political, and transformational analysis in promoting social and political change over the boundaries of nation states and the institutional forces seeking to establish a neoliberal matrix of individualization and power. We have shown that other forms of sociability across state borders are possible and with this comes the organization, agency, and creativity of quotidian materialities that stimulate the hegemonic forces of capitalism to reconstruct themselves, establishing new relations of intimacy, intra-subjectivities, and affect through stimulating the emergence of innovative public engagements that have the potential to create new ways of understanding and changing the world.

Acknowledgment

This investigation was sponsored by the Consejo Nacional de Ciencia y Tecnología (CONACYT-México).

Notes

1 Ethnographic accounts and interview material used in this text are drawn from the doctoral thesis of Díaz Copado (2013). Methods for data collection on the chili cook-off include a review of organizational material available on the internet and Díaz undertaking participant observation as a volunteer at the three-day cook-off competition in February 2008, plus interviewing participants and attending wider charitable events in the community between 2007 and 2008. All names are pseudonyms.

2 Information on the UK Chilli Cook-off Association is available at: www.ukchillicookoff.co.uk. The Reading Chilli Festival is an example of a local event with information available at: www.readingchillifest.co.uk/home.html (both websites accessed 11 August 2016).

3 The webpage of the Mexican National Chili Cook-off in Ajijic, Mexico – notable for being in English without option for Spanish translation – is available at: www.mexicannationalchilicookoff.com/index.html. Accounts are also available at: www.accesslakechapala.com/2015/03/03/37th-annual-chili-cook-off-in-ajijic/ and www.facebook.com/MexNatChCookOff (both websites accessed 11 August 2016).

4 In 2016, seven charities were supported by the cook-off: Escuela para Niños Especiales; Centro de Desarrollo (Ajijic); Casa de Ancianos de Chapala / Home for Elderly Citizens in Chapala; Cruz Roja de Chapala / The Red Cross of Chapala; Love In Action Children's Home; Niños Incapacitados del Lago; and Los Niños de Chapala y Ajijic, A.C. (NCA). Available at: www.mexicannationalchilicookoff.com/MexicanNationalChiliCharities.html (accessed 11 August 2016).

5 For pictures see: www.accesslakechapala.com (accessed 11 August 2016).

6 A pre-Colombian ritual reproduced by four natives of Ajijic since 2007.

References

Arce, A. (2010). State policy intervention in an era of civic participation. In N. Long, Ye Jingzhong and W. Yihuan (eds) *Rural Transformation and Development – China in Context: The Everyday Lives of Policies and Peoples*. Cheltenham: Edward Elgar, 279–311.

Arce, A. and Long, N. (2000). Reconfiguring modernity and development from an anthropological perspective. In Arce, A. and Long, N. (eds), *Anthropology, Development and Modernities: Exploring Discourses, Counter-Tendencies and Violence*, 1st ed. London and New York: Routledge, 1–31.

Bennett, J. (2010). *Vibrant Matter: A Political Ecology of Things*. Durham, NC: Duke University Press.

Blanco, G., Arce, A. and Fisher, E. (2015). Becoming a region, becoming global, becoming imperceptible: territorializing salmon in Chilean Patagonia. *Journal of Rural Studies*, 42, 179–190.

Braidotti, R. (2013). *The Posthuman*. Cambridge: Polity Press.

Braidotti, R. (2015). Keynote Lecture – Posthumanism and Society Conference, New York 9 May 2015. Available at: https//youtu.be/3S3CulNbQ1M (accessed 17 August 2016).

Carolan, M.S. (2011). *Embodied Food Politics*. Farnham: Ashgate.

Csordas, T.J. (1990). Embodiment as a paradigm for anthropology. *Ethos*, 18(1), 5–47.

Csordas, T.J. (ed.) (1994). *Embodiments and Experience: The Existential Ground of Culture and Self*. Cambridge: Cambridge University Press.

Csordas, T.J. (1999). Embodiment and cultural phenomenology. In Gail Weiss and Honi Fern Haber (eds), *Perspectives on Embodiment: The Intersections of Nature and Culture*. Abingdon: Routledge, 143–62.

De Landa, M. (2006). *A New Philosophy of Society: Assemblage Theory and Social Complexity*. New York: Continuum.

Deleuze, G. and Guattari, F. (1987). *A Thousand Plateaus: Capitalism and Schizophrenia*. London: Athlone Press.

Dewey, J. (1954). *The Public and its Problems*, 2nd ed. Athens, GA: Swallow Press.

Díaz Copado, F.V. (2013). *Shaping Multiple Ajijics and Development. A Mexican Town in the Context of the International Retirement Migration*. PhD dissertation, Wageningen University, The Netherlands.

Gabrielson, T. and Parody, K. (2010). Corporeal citizenship: rethinking green citizenship through the body. *Environmental Politics*, 19(3), 374–391.

Guattari, F. (2006). *Chaosmosis: An Ethico-Aesthetic Paradigm*. Sydney: Power Publications.

Hayes-Conroy, A. and Martin, D.G. (2010). Mobilising bodies: visceral identification in the Slow Food movement. *Transactions of the Institute of British Geographers*, 35, 269–281.

Ingold, T. (2011). *Being Alive: Essays on Movement, Knowledge and Description*. London and New York: Routledge.

Krarup, M. and Blok, A. (2011). Unfolding the social: quasi-actants, virtual theory, and the new empiricism of Bruno Latour. *The Sociological Review*, 59(1), 42–63.

Long, N. and Long, A. (1992). *Battlefields of Knowledge: The Interlocking of Theory and Practice in Social Research and Development*. London and New York: Routledge.

Müller, M. and Schurr, C. (2016). Assemblage thinking and Actor–Network Theory: conjunctions, disjunctions, cross-fertilizations. *Transactions of the Institute of British Geographers*, 41(3), 1–13.

Ritchey, B. (2007). The history of chili cook-offs. In Raven J. (eds), *Part One: In the Beginning*. Available at: www.texascooking.com/features/march2007_chili_cook-offs.htm (accessed 17 August 2016).

Tsing, A.L. (2015). *The Mushroom at the End of the World: On the Possibility of Life in Capitalist Ruins*. Princeton, NJ and Oxford: Princeton University Press.

Van der Tuin, L. and Dolphijn, R. (2010). The transversality of new materialism. *Women and Cultural Review*, 21(2), 153–714.

2

CREATING HEALTHY BODIES IN RURAL ECUADOR AT A TIME OF DIETARY SHIFT

Joan Gross, Carla Guerrón Montero, Michaela Hammer, and Peter Berti

Introduction

Human bodies are not sovereign. We begin our lives dependent on the adults who care for us, and we continue being dependent on other people and on our environment, which provides sustenance, but also toxins. People have developed and continue to develop ways to keep bodies healthy in order to ward off possible problems, acknowledging that our equilibrium is precarious. This chapter is a foray into how people act to keep themselves and their loved ones healthy in peasant communities in northern Ecuador. Our particular focus was on the eating and feeding practices of mothers of young children since the emergent quality of the embodiment of food practices can be seen most clearly in the visible growth of young children. We soon realized that curing practices were seen as integral to the creation of healthy bodies and they exhibited the same frictions between knowledge (practices) passed down through families and the advice of health professionals. Embodied practices of feeding oneself and one's children have a habitual element to them, yet they are subjected to new contexts as farming communities are brought into global modern markets replete with new foods and new constraints on time and money.

In this chapter, we follow Michael Carolan's (2011) lead to think about the embodiment of food politics as well as the relations that hold people together through food. In doing so, we situate flavor, taste, and eating within layers of production in the fields and in the kitchen. The empirical study of situated food practice sheds light on how transformation occurs without simplifying its heterogeneity. Over the past several decades, a global dietary shift toward ultra-processed food that is more energy dense and less nutrient dense[1] has been occurring alongside a rise in non-communicable chronic diseases and the rise of pandemic overweight/obesity (Monteiro *et al.*, 2012; Popkin, 2004). The shifts associated with these processes are commonly referred to as the "nutrition transition," and have been largely characterized by an increase in consumption of inexpensive edible oils, sweeteners, and animal-sourced foods alongside a reduction in the consumption of nutrient-dense legumes, whole grains, fruits, and vegetables, all significant parts of traditional diets (Drewnowski and Popkin, 1997; Popkin *et al.*, 2012). Rural areas of Ecuador are beginning to exhibit dietary-related embodiments of modern food, but not at the same rates of urban populations where ultra-processed foods and supermarkets are more readily available and obesity rates are at their highest (Freire *et al.*, 2013).[2] Rural areas like the one we focus on still eat a mainly whole foods diet. They produce a fair amount of their own food and buy additional food at traditional open-air markets and small retail stores where they can access credit.

This chapter draws on experiences in two highland Andean villages near the towns of San Isidro and El Angel (approximately 3,200 meters above sea level (masl)) and a lowland area around the town of Ambuquí (about 1,860 masl). Family farming is prevalent in these two regions, despite a strong presence of dairy and potato industrial production at large and medium scale in the highlands and large-scale sugar cane production in the valley. In the two higher villages, potato–pasture systems of production dominate the landscape, while in the valley fruit and beans are the major products. Families are connected to urban markets in their role as buyers of agrochemicals and as sellers of vegetables, fruit and milk as well as the purchase of other basic goods. Some villagers also perform waged labor, both in the agricultural and non-agricultural sectors. People commonly watch TV, listen to the radio and have access to public services, including local health centers. Through all of these sources, families encounter different information on food products, health and nutrition.

We start from the premise that transformations such as the nutrition transition are never unidirectional, but rather complex struggles "back and forth, up and down and back again" between multiple actors, scales, and levels of presumed authority allowing us to construct a "history of the present" (Wright and Reinhold, 2011: 101). We predominantly used an ethnographic approach to access the messiness of the ways in which old and new ways of producing and consuming food are coming together and co-creating different kinds of foods and bodies. Preliminary fieldwork was done in the fall of 2011, then from March to August 2013, four ethnographers[3] lived in the villages, spending from three to seven days with 25 families, conducting participant observation and noting discourses about food, food procurement, cooking techniques, meal construction, child feeding practices and hygiene in the home. The ethnographers ate meals with participating families and engaged in informal conversations with all those present. They also conducted semi-structured, recorded interviews with the mother of each family.[4] These interviews addressed dietary practices during pregnancy and lactation, household food security, family food production, home remedies and health-care practices. We were particularly interested in how new materialities, such as ultra-processed food and agrochemicals (as well as roads and vehicles) shape relationships between human bodies and the substances and activities that create them.

Diet diversity in context

Ecuador's diverse altitudes contribute to diet diversity and encourage reciprocal relations between unique ecological zones (Camacho, 2006). Exploiting land at different altitudes and participating in reciprocal trade with people who have access to other products lay at the basis of Murra's (1967) model of verticality for the pre-Columbian Andean region. We found clear patterns of exchange of food produced in high and low altitudes in these contemporary villages. For instance, the people of Ambuquí trade with friends and family at higher altitudes for grains and they supply fruit and beans in return. As Hoogesteger notes, "bonding" and "bridging" ties are fundamental aspects of inward- and outward-looking relationships that generate and facilitate trust and reciprocity among Andean populations (2013: 349).

Twenty-four hour dietary recalls with mothers in 140 households in the three villages provided a snapshot of the diverse ingredients consumed in a single day by a variety of people. As in most agrarian societies, meals are composed of a starchy core complemented by a fringe of other foods (Mintz and Schlettwein-Gsell, 2001). The highland villagers are known as potato eaters after their common starchy core. In the dietary recalls, 100 percent of the people in one village and 97 percent in the other ate potatoes on the recorded day. People in Ambuquí also reported high consumption of potatoes (84 percent), but they were more likely to follow the national predilection for refined white rice.

The dietary recalls also showed individual variation. Take, for example, María who ate only 12 ingredients during the 24 hours recorded and Lucía who reported 52. When examining exactly what was eaten at what time, we noted two very different patterns of eating. María drank a glass of Tang with bread at 9 a.m. She had to skip lunch because she was working for a wage, but paused at 2 p.m. to have a soda with bread. She had fava beans and soup (potatoes, milk, egg, onion, annatto and a seasoning packet) for supper. Lucía had coffee with sugar at 7:30 a.m, along with potato soup (egg, onion, annatto, and milk), a cooked egg, and a homemade wheat tortilla. At 1:00 p.m. she ate fava beans mashed with onions, salt, lard, and annatto; potatoes, rice, and homemade naranjilla juice. At 5:00 p.m. she had coffee and tortillas as a snack and at 8:00 p.m. she had potatoes and soup (noodles, onion, lard, milk and cilantro). Both María and Lucía are married to agricultural laborers and reported that some of their ingredients (María – three and Lucía – ten) came from home production, but Lucía's schedule allows her more time to cook and eat. María seems to have already tipped into a modern food regime, consuming bought bread instead of homemade tortillas, instant instead of homemade juice and processed seasoning packets in her soup. As is usually the case with ultra-processed foods, all these products saved her time, but reduced the healthfulness of her diet.

Diets also change throughout the year based on religious calendars, life cycle events and when particular crops ripen. For instance, while few people eat meat daily, it is an important centerpiece of celebratory feasts. In the two highland villages, women raise pigs to be slaughtered for baptisms, first communions, confirm-ations, graduations, weddings, and other life cycle events. The pig is marinated for

24 hours in a sauce containing cilantro, parsley, salt, onion, garlic, cumin, pepper, tomato, oregano and celery and then roasted for another 24 hours in a wood-burning oven. In Ambuquí, the meat is more likely to be guinea pig or chicken but in both highland and lowland villages copious amounts of food are given to attendees and they each take their leftovers home with them. Celebratory foods are to be shared with others to cement relations between individuals and families.

Embodying new materialities

Estela and Marta said that they used to walk everywhere. Marta was born in a house outside the village and had to walk down to the *acequia* (canal) every morning to get water for the house and to wash clothes. Now there is a road to the village and piped-in water. They used to walk two hours to get to the El Angel market, three hours to get to Mira. Children started walking long distances early with their mothers. Estela used to walk with her two-year-old son to meet his father who worked as a guard at the nature reserve. The trip took an hour and a half and rose in altitude from about 10,000 to 12,000 ft. Now there are so many vehicles on the roads that the long walks of the past are over. Even if someone starts out walking, they will likely be picked up before they reach their destination. Pickup trucks are especially nice because so many people can fit in the back. Even when the driver has errands to run along the way, passengers sit patiently in the back, perhaps remembering the hours it took them to walk in the old days.

Roads and cars changed the physical activity level of the population, but they also changed the food economy because with roads came market integration. Vehicles came from outside buying what was produced locally and selling what was not. Now, there are trucks that come once or twice a week selling fruits and vegetables and a car with fish and shrimp that usually comes once a week. The local stores are also filled with white rice, candy of all kinds, and other highly processed foods that come in on the roads. Roads and vehicles also allowed people better access to external markets for their own production. Carchi province has been supplying most of the nation with potatoes since the 1980s. With roads, they could also sell milk every day, providing a steady source of income. No fewer than five trucks from different companies come through the highland towns every day collecting the milk that came from the cows early that morning, or at the earliest, the previous afternoon.

Entering the market economy and its accompanying technological regime brought farmers in Carchi into the modern age of agriculture dominated by the agrochemical companies whose murals adorn multiple walls of buildings and stadiums around the region. Families have learned to keep their pesticide storage away from their small children, but that does not stop anguished teenagers from occasionally ingesting agrochemicals to end their lives. Farmers saw their yields and incomes go up with modern chemical inputs, but their good fortune did not last. At the time of our fieldwork, the price of a quintal of potatoes had dropped below what it cost to produce it. Farmers sometimes received up to US$ 35 per quintal

(100 lbs.), but recently the price had dropped to US$ 4, even though the retail price of potatoes in Quito continued to hover around US$ 50.

For some, market integration began to seem more like a contract with the devil, like the local legend of the construction of the bridge. This legend details how the bridge collapsed into the deep ravine numerous times before completion. Then, a construction worker drove his truck off the unfinished bridge, plunging to his death, and the bridge was completed. After that other people ended their own lives by leaping off. People began to wonder if these lives were taken by the devil in exchange for allowing the bridge's completion. If so, has the quota been reached or not? Some people still try to avoid the bridge, but it is the fastest route to the market town, interestingly called Angel. Market integration is full of risks.

Entrance into the milk market changed consumption patterns in the home. Marta talked about how much milk they used to use: they cooked rice, *morocho* and *mote de trigo* in pure milk. They made cheese and added big chunks of it to rice and potatoes. They used the whey to feed their pigs. Sometimes they even sold cheese to neighbors without cows. Nowadays, people sell their milk to one of the companies that buys from small producers. People hardly use their own milk anymore, said Marta. They keep some back every day for coffee or to put in soups, but practically the only time people make cheese is during *Semana Santa* (Holy Week) when the cheese factories are not operating.

When suggested to Marta that people like the money more than the milk, she strongly rejected that interpretation with the following response:

> The people from here don't see the money. When harvest comes, we harvest potatoes and barley and then we have something to sell, but other things take the money, fertilizer, pesticides and herbicides, machines, the harvester itself, now we have to pay for the harvester for the barley. It's as if the money comes and just passes through. Now most people hand over their milk and there they have enough for the week to buy salt, lard, soap, sugar and anything else ... This is life in the country.
>
> *(24 July 2013)*

Marta lamented the way that money slips into and out of people's hands in the countryside with nothing to show for it. Money that comes in has already been spent since stores regularly allow people to buy on credit. "Modern" farming devoured all the extra income generated by selling milk, and women began working in the flower industry or contract knitting in order to earn extra money.[5] This gave them even less time for supplying their households with food.

Milk and healthy beginnings

Despite the incentives to sell cow's milk rather than feeding it to their families, milk is the second most common animal-sourced food (after eggs) in our dietary survey. It is regularly added to soups and coffee and sometimes to fruit juices and coladas.

Drinking a cup of warm milk is considered a pleasure in the region, though opinion is divided as to whether it is best raw or boiled. Having a cow and being able to milk gives one access to the milk foam that they prize for its deliciousness. After the cows have had their calves, the milkers have access to colostrum, which they prize for its health-giving properties.

Turning to health leads us to human milk. The importance of this first food is encoded in cultural practices that support the health of women. During preliminary research, we held meetings in each village and asked how healthy children are nurtured. We were told that the first step is having a well-fed mother. The mothers we interviewed later recounted that during their pregnancies they were encouraged to eat whatever they wished and whenever they were hungry. Particular significance was given to the time when the bodies of mothers and babies are first separated, yet still connected through the milk that flows out of the mother and into the baby. The cultural practice of *la dieta* (the diet) facilitates this healthy beginning. After a child is born, the mother is cared for so that she is able to produce enough milk to create a healthy child. New mothers are prevented from doing physical labor and fed by close relatives, usually their mothers. There was general agreement that traditionally this lasted for 45 days, but that the period had been shrinking in recent times. The main concern is to keep the woman warm because cold is thought to dry up her milk. This means a considerable reduction of her normal duties. New mothers are prevented from going outside and doing anything with water – no washing, no peeling potatoes, no bathing. Soup made from free range roosters is the most common element of the diet, but also various warm drinks made of ground oats, barley or corn (*coladas*), herbal teas, and fruit juice. Both the coladas and fruit juices are often made with milk. It is commonly believed that consuming warm liquid leads to milk production, but contact with cold liquids does the opposite. Potatoes, the most common food in daily life, are severely restricted, especially in the evening. Some people said that potatoes irritated the mother's stomach and others said that they irritated the baby, at least in the first few weeks. White carrots[6] are mashed instead of potatoes or added to the soup. Mothers also mentioned chard, mustard greens, and carrots as vegetables that they add to the chicken soup along with *mellocos*, commonly understood to heal the wounds of childbirth. Beans, one of the main sources of protein in the local diet, are avoided and replaced by higher quality meat.[7] A concern about fat led some people to exclude pork (the most common meat) from *la dieta*.

All of the mothers we observed and interviewed breastfed their children. (One woman said that if a woman does not breastfeed her baby, the milk solidifies in her breasts and causes cancer.) Only one of the women we interviewed practiced exclusive breastfeeding for the first six months (the WHO recommendation), while the others supplemented breast milk with formula, teas, *coladas*, and soups. Women nursed their babies on demand rather than on a specific schedule and weaned them from 9 to 26 months. Some children abandoned the breast easily, while others were difficult to wean. In these cases, mothers painted their nipples with nail polish, hot pepper sauce, coffee, or astringent latex from a local flower to discourage the baby from suckling.

Mothers supplemented their breast milk with other liquids in the first six months of infancy for a variety of reasons. It was common to start giving infants very small amounts of medicinal teas, such as oregano, chamomile, eucalyptus and aniseed, sometimes made with refined cane sugar. These were sometimes introduced alongside breast milk soon after birth, as observed in one family with a three-week-old daughter who was given sips of tea from a spoon. This was said to maintain stomach health and to strengthen the infant's digestive system before they introduced more substantial liquids and foods. Some mothers did not have enough milk to satisfy their babies. The trauma of a Caesarian section can delay breast-feeding and a reduction of breastfeeding time reduces the amount of milk produced, since frequency of consumption stimulates production. One mother attended school in another province and could only breastfeed when she came home in the evenings. During the day, her mother would feed her son with either pumped breast milk or formula milk. Another mother worked full time away from home. Her sister-in-law who also had a baby looked after her niece who she fed either with formula milk or at her breast. These two sisters-in-law both weaned their daughters by giving them four or five packets of sweetened, drinkable yogurt to suck throughout the night. We found this type of store-bought yogurt to be a common first food for babies.

Complementary feeding of solid foods began for the majority of mothers at six months, but sometimes it began as early as four or as late as nine months. We were told that breast milk has many vitamins and nutrients for the first six months, and then it becomes like any other food – nothing special, just sustenance. Besides herbal infusions and yogurt, first foods included *coladas*, soups, grated apple, mashed bananas and potatoes, whole fruits to suck on, wild mustard greens, and eggs. Mothers introduced these to their babies gradually, progressing from simpler liquefied foods, such as oatmeal or maize coladas, to more complex, solid foods, such as rice soup or scrambled eggs with vegetables. Seasonings and colorants were used sparingly in first foods. While it was common practice for mothers to use small amounts of what the rest of the family ate, some mothers prepared special dishes for their babies. Mothers fed their small children every two hours, starting with the breast if they were still breastfeeding. They said that their stomachs were small and they could not wait like adults. Sometimes they had to resort to games and threats to get their children to eat. Mothers were deeply affected by their children. They were happy when they saw the conversion of food into child growth and worried when children were thin and their growth stagnated.

Preventing and curing illness[8]

Mothers fostered a variety of relationships with humans and nonhumans in order to keep their children healthy. They took their children to the local health center regularly to get weighed and vaccinated, but when there was disagreement between the advice of health professionals and village health traditions, mothers tended to prioritize family practice. For instance, one mother was told by a doctor not to give her baby citrus for the first year, but the mother gave her ten-month-old baby mandarins and oranges because it was common practice and her baby was developing well. In this same spirit, in the event of illness, mothers typically applied home remedies prior to consulting a medical professional.

The most common child illnesses among participating families were respiratory disease (including cough, cold, sore throat, bronchitis, and pneumonia), gastro-intestinal disorders (including diarrheal infections, stomach aches, constipation and rotovirus), and *espanto* or *mal aire*. Mothers tried to prevent respiratory problems by protecting children from cold temperatures (layering clothes and keeping a fire lit) and germs (coughing into a handkerchief, washing hands, keeping distance). They also gave orange juice, sometimes freshly squeezed and sometimes boiled with cinnamon and sugar. Other remedies included garlic, honey, mandarin peel, wild cherries and soups, all of which tended to be prepared with boiled milk. In addition, they gave herbal teas based on a selection of plants gathered around the house: eucalyptus, chamomile and elderflower.

Teas were even more common in the prevention and treatment of gastro-intestinal troubles. Mothers explained that it was important to treat digestive complaints immediately with teas, so that the illness did not progress into something worse. Good hygiene was seen as important in preventing this kind of illness and was often linked to food. There was a particular fear of re-heated foods and foods that were not well cooked. Even if adults could withstand re-heated foods, the small stomachs of young children could be quickly overwhelmed. The most common remedies for gastrointestinal illnesses included onion, cumin, barley gruel, chamomile, mint, oregano, granadilla, anise, guava and thyme. One mother started her baby on oatmeal, but added some oregano leaves "so that if his little tummy wants to hurt, it already comes with the oregano, so it doesn't hurt him." This desire to prevent harm to the bodies of loved ones was echoed in other cooking practices. One mother placed a long onion in every pot of rice she cooked, removing it before serving, explaining that it helped with digestion. Medicinal food practices become normalized as daily habits, melding seamlessly into the mundane.

If home remedies were not working with respiratory or gastrointestinal illnesses, medical doctors were consulted, but the third category of illness lay completely outside the domain of biomedicine. *Espanto* frequently befell babies and children after becoming frightened. The symptoms included fever, cold feet, vomiting, lack of energy, no appetite, not being able to sleep or sleeping too much. Mothers feared the more serious *pasmo/mal aire* if their children developed these symptoms after being in the countryside. *Pasmo/mal aire* must be treated quickly a certain number of times and on particular days. We were told that bad spirits resided in the clouds and wild areas and that entities like rocks and water could capture the souls of vulnerable people. This could have lasting effects. One boy fell into an irrigation canal as a small child and even though he went through a rigorous cure, his grandmother said that he still was thin and had low energy years later. They explained that children's souls are not fully attached to their bodies, so they could easily be snatched or simply left behind. That is why they said it was important to hug babies and to call your children by name in order to prevent bad energies from taking hold of them.

The treatment for these illnesses involved human-to-human contact through the mediation of nonhuman substances. The healer (who might be the mother, another relative or a local expert) swept branches over the patient's body, pressed or rubbed oil into parts of the body, blew tobacco smoke and spat alcohol on the patient. Special words in an unrecognized language were pronounced at particular times. Sometimes eggs or guinea pigs were passed over the patient's body and they assumed the illness, removing it from the human. One mother closed the curtains so that her child could not look at clouds and forest, the source of the problem. Foods were not seen to prevent or cure these illnesses, but during the three days of curing, the patient had to avoid eggs and dairy products – the two most common animal-sourced foods.[9]

Harmonious relations within the family were cited as important for creating healthy children. In the case of mothers calling their children by name to prevent their soul from being left behind, we see a specific example of how expressed relationships actually protect the child. One author watched one mother feeding her baby and saying the names of each family member loudly and clearly between each bite, encouraging the baby to mimic her. Children were taught to participate in binding the community together through the linguistic practice of greeting. One always greeted another when passing and greetings were shouted out at some distance when approaching a house.

Just as good relations were nurtured in the villages through greeting, good relations with plants and animals also had to be maintained. Caring for these domesticated beings defined life in the villages. Care grows from knowing the specific qualities of plants: what they need from you and what they can give to you. Knowledge about the relationships between human bodies and the plants and animals that provide nourishment for those bodies was seen as a tool with which women create healthy children. This knowledge was passed down mainly from mother to daughter over the course of a life of careful observation of family food practices. Knowing the qualities of herbal teas was an important domain as noted above. Common herbs were mentioned 33 times in our dietary records and 16 different herbs were mentioned at least once. Other plants have more human-like characteristics, like the flower on the *páramo* that closes up if you speak the truth[10] or a certain tree that makes your head and hands swell up and itch if you neglect to greet it when you pass.

Localized modernities

When people talked about how to handle vindictive trees, *espanto*, and *pasmo/mal aire*, they indicated an understanding that these practices were not modern. One mother said that she only partially believed in what she was saying, but nonetheless it was still part of her practice. This reminds us of the messiness of social transformation and the back-and-forth struggles as the intimate relationships we have with foods and curing objects shift. Even within what at first glance may appear to be a homogeneous community of peasant farmers, families create their own heterogeneous patterns of preparing foods and eating them. At different moments, each partakes in store-bought, processed foods, while also reproducing practices from earlier generations. The importance of traditional or new ways of eating is nuanced and varies from family to family, thereby shaping multiple assemblages of modernity.

Young and old alike spoke of people in the past being healthier due to healthier diets and agricultural practices. It was commonly stated that previous generations lived much longer. Cristina, a 37-year-old mother, said that nowadays 30-year-olds suffer from maladies that used to be restricted to much older people. She added that instead of sweets, older generations ate toasted fava beans and toasted corn, and instead of bread they drank máchica. "They also didn't spray their crops [with pesticides]," she added. She explained that barley and corn were healthier because they were "born of the earth," not like cheetos and cookies that were loaded with "sugar and fat." The double-voicedness in her statement is striking, as it uses a quasi-spiritual metaphor when describing whole foods, while breaking ultra-processed foods into macro-nutrients.

Grandparents in the participating families asserted that refined commercial foods, such as store-bought bread, white rice, soda, pasta, and white sugar were unavailable when they were young. They spoke of growing up eating primarily locally produced foods, such as *mote de trigo* (wheat berries), *arroz de cebada* (barley gruel), *máchica* (a toasted barley-based drink), *habas* (fava beans), *tostados* (toasted corn), *morocho* (often in the form of a thick, sweetened corn beverage), and, of course, different varieties of potato. Many of these food sources required drying, toasting, grinding and cooking for long periods. One might say that they forced an ongoing relationship on the people they fed. Foods that one buys today in the grocery store are not as demanding, but there is a sense that they do not return as much either.

Margarita, a mother of two young children, assigned the cause of the dietary shift on women's desire for less work and the fussy eating habits of children. In the following quotation, we get a sense of the different circulating discourses that influence food choice. She alluded to an assortment of absent interlocutors, represented by "they." There are the "theys" of "before" – the agents that did the work to make these old foods palatable. Then there are the "people nowadays" who only want food that is easy to prepare. At the same time, there is a contrasting group of contemporary interlocutors who point out the health-giving attributes of traditional foods, such as *mellocos*. We are left to wonder who the "But now they say" refers to in the final sentence of this explanation. These actors appear in the guise of experts, but the tradition from which they come is unclear. People may hear discourses about the medicinal value of traditional crops from alternative or conventional doctors, from new age health programs found on the radio, or in the marketplace.

> Before they planted a lot of barley in Chitacaspi; now only potatoes. Before, barley was like potatoes are today; you couldn't have a meal without barley. Before they used to make *chapo*[11]: a kind of *pinol* made with dried, toasted barley that was ground. They called it *aco*. To toast and grind the barley was work. People nowadays don't like anything that takes a lot of work. They only want [pre-prepared] tuna, noodles, whatever's easy. Kids today ask what kind of soup it is before they eat it. They don't eat savory *arroz de cebada*; they're used to noodle soup, noodles or rice. Before they ate *ocas* with milk, even *mashua*. I didn't like it. I had to carry them and when no one was looking, I threw them out. There were also *mellocos* of various colors, pink, yellow, red. I liked the red one best, because it wasn't slimy. But now they say that the slime of the *melloco* is good for bones – that it's like oil between bones, so that they won't break.
>
> *(27 September 2011)*

Unlike some of the people we talked with, Margarita does not miss the taste of traditional foods. She even admits to throwing out the tubers that she did not like, but she is convinced that they were healthier than today's convenience foods. People often noted that life in the past was more difficult and they were not interested in returning to a time without electricity and piped-in water. However, they saw a connection between modern foods and poor health and this led them to emphasize the health-giving qualities of traditional foods, mitigating a rapid dietary transition.

But the pull of participating in contemporary foodways is strong. People who lived in the study area might have rarely or never visited the large cities of Quito and Guayaquil. Nonetheless, they are targets of competing social realities and their ideologies emanating from urban centers. New financial practices, government health policies, and proposed lifestyles are embedded in advertising campaigns intended to gain influence over peoples' eating habits. They see advertisements for the latest snacks and convenience foods and for fortified Nestlé products on television. The lifestyle promoted in these ads stands in stark contrast to the daily life of rural food producers, and the money needed to obtain advertised products is usually scarce, but the desire to be a modern Ecuadorean is still nurtured, especially among the youth whose identity is commonly less anchored in the reproduction of tradition.

In addition to the arrival of a globalized food industry, international tastes seeking fresh vegetables year round have led to local farmers planting new crops and some of these find their way into the farmer's kitchen. Some local women were starting to serve broccoli, chard, lettuce, and cauliflower according to recipes that they saw on television. This change in culinary practice was not always appreciated by relatives who complained that eating leafy greens is "eating like an animal." Other new foodways were brought home directly by "vernacular cosmopolitans" who went elsewhere, discovered new flavors, and learned new cooking techniques (Werbner, 2008: 14). Women who worked in the kitchens of haciendas, rich urbanites, or urban restaurants learned "fancier" ways of cooking and taught them to their daughters, who appropriated them with enthusiasm. One family in Ambuquí enjoyed how tasty their soups were thanks to a secret learned by Gioconda, the eldest daughter in the family, who worked in a restaurant owned by former migrants to Spain. She showed us how she would peel a large onion, add a carrot and parsley and sautée these ingredients with oil and annatto seed. Then, she blended this mixture and added it directly to the soup. Culinary shifts are seen as improvements by some, but others do not appreciate such innovations. One woman was chastised by her husband for "cooking like a rich person."

Conclusion

Empirical food studies show that food practices collapse classic dichotomies dividing traditional from modern, local from international, and urban from rural practices.

In our study, we learned that just as memories of the past were both positive and negative, so too were opinions about cosmopolitan food practices. The global nutrition transition may be profoundly affecting the general diet of Ecuadoreans, but not uniformly. In the villages we studied, we found heterogeneous nutritional patterns motivated by discourses of tradition and modernity, experiences of illness and desires for particular tastes, all within a context of seduction into a cash economy that commonly disadvantages traditional peasant agriculture.[12] In the midst of this, women grapple with information coming from their older relatives, medical doctors, and popular culture. They learn about earlier foods that demand attention, new foods that free up time, and the qualities of different herbs. Within this complex network of relationships, they create new humans who eventually will construct their own practices within hybrid assemblages that are urban and rural, local and international, traditional and modern. Gibson-Graham refers to this as a "process of becoming" (2006: 25), where individuals are able to imagine alternative subjectivities and craft new livelihoods and identities, opening up possibilities for different kinds of food systems in the future.

Acknowledgements

The research reported here was part of a Dutch-financed NWO–WOTRO Programme on "Catalyzing Positive Deviance in Agriculture and Food."

Notes

1 That is, nutrients per gram of food, or per kilocalorie of food energy.
2 Ultra-processed foods in particular, argue Monteiro *et al.* (2012), are highly profitable

and are being vigorously marketed in the Global South, where sales are increasing by at least 10 percent each year (Lacroix *et al.*, 2013).

3 Michelle Soto did preliminary research, followed by ethnography by Gabriela Jaramillo and the first three authors. A fourth author, Peter Berti, supervised the dietary recalls.

4 Ethics approval for this study was granted by the Institutional Review Boards of Oregon State University (#5704) and the Universidad San Francisco de Quito (#2012-68). We obtained verbal and written consent from adults and verbal assent from children. All names used in this article are pseudonyms.

5 Many women in the highlands have wool delivered to them by an intermediary, along with orders for hats and sweaters. They spend all their spare time knitting to bring extra money into the household, earning 50 cents for a hat that sells for at least US$ 5 at the market in Otavalo.

6 *Arracacia xanthorrhiza* is an Andean root, different from carrots, which is white.

7 When older women suffer from joint pain and arthritis, people said that it was because they were not taken care of properly during *la dieta*. Meat consumption is seen as particularly important.

8 Much of the information in this section was first presented in Hammer (2014).

9 As in *la dieta*, we see once again the avoidance of common foods during liminal phases.

10 A doctor told one pregnant mother that the baby she was carrying was a boy, but when she asked the flower, it didn't close. When her daughter was born everyone was surprised except her.

11 *Chapo* is another word for *máchica*.

12 By traditional peasant agriculture we refer to diversified farms using animal manure rather than chemical amendments (see Gangotena, 1981).

References

Camacho, J. (2006). Good to eat, good to think: food culture and biodiversity in Cotacachi. In R. Rhoades (ed.), *Development with Identity: Community, Culture and Sustainability in the Andes*. Oxford and Cambridge, MA: CABI Publishing, 156–170.

Carolan, M. (2011). *Embodied Food Politics*. Burlington, VT: Ashgate.

Drewnowski, A. and Popkin, B. (1997). The nutrition transition: new trends in the global diet. *Nutrition Reviews*, 55(2): 31–43.

Freire, W.B., Ramírez, M.J., Belmont, P., Mendieta, M.J., Silva, K., Romero, N., Sáenz, K., Piñeiros, P., Gómez, L.F., and Monge, R. (2013). *Resumen Ejecutivo, Tomo I*. Quito: ENSANUT-ECU 2011-2013/Ministerio de Salud Pública/Instituto Nacional de Estadística y Censos.

Gangotena, F. (1981). *Peasant Social Articulation and Surplus Transference: An Ecuadorian Case*. PhD dissertation, University of Florida.

Gibson-Graham, J.K. (2006). *A Postcapitalist Politics*. Minneapolis, MN: University of Minnesota Press.

Hammer, M. (2014). *Hybrid Healthscapes: Mothers' Health Care and Food Practices in the Ecuadorian Andes*. MA thesis, Oregon State University.

Hoogesteger, J. (2013). Social capital in water user organizations of the Ecuadorean Andes. *Human Organization*, 72(4): 347–357.

Lacroix, P., Chauveau, C., and Taipe, D. (2013). Soberanía alimentaria y comercialización campesina en Ecuador. In F. Hidalgo, P. Lacroix and P. Roman (eds), *Comercialización y soberanía alimentaria*, 1st ed. Quito: Sistema de investigación sobre la problemática agraria en el Ecuador (SIPAE), 63–68.

Mintz, S. and Schlettwein-Gsell, D. (2001). Food patterns in agrarian societies: the core-fringe-legume hypothesis – a dialogue. *Gastronomica: The Journal of Food and Culture*, 1(3, Summer), 40–52.

Monteiro, C., Cannon, G., Levy, R.B., Claro, R., and Moubarac, J.C. (2012). Ultra-processing: the big issue for nutrition, disease, health, well-being. [Commentary]. *World Nutrition*, 3(12): 527–569.

Murra, J. (1967). La visita de los Chupacu como fuente etnológica. In J. Murra (ed.), *Visita de la provincia de León de Huánuco en 1562*. Vol. 1. Huánuco, Peru: Universidad Nacional Hermilio Valdizán, Facultad de Letras y Educación, 383–417.

Popkin, B.M. (2004). The nutrition transition: an overview of world patterns of change. *Nutrition Reviews*, 62: 140–143.

Popkin, B.M., Adair, L.S., and Ng, S.W. (2012). Global nutrition transition and the pandemic of obesity in developing countries. *Nutrition Reviews*, 70(1): 3–21.

Werbner, P. (ed.). (2008). *Anthropology and the New Cosmopolitanism: Rooted, Feminist and Vernacular Perspectives*. Oxford: Berg.

Wright, S. and Reinholt, S. (2011). Studying through: a strategy for studying political transformations. Or sex, lies and British politics. In C. Shore, S. Wright, and D. Però, *Policy Worlds: Anthropology and the Analysis of Contemporary Power*. New York: Berghahn, 86–104.

3

FINDING THE FOOD BY HIDING THE GOLD

Andoque abundance, mining, and food in the Colombian Amazon

Camilo Torres and Gerard Verschoor

Introduction

Food is important – but so is thinking about food through knowledge and experience. Food is an essential part of the "web of life," but thinking about food has real consequences for the way it is grown, gathered, or procured. Scholarly and common ways of "thinking food" help shape systems of food production, distribution, and consumption, which have a bearing on the way societies are organized. As Carolan (2013: 413) notes, academic ways of thinking about food have undergone a number of "turns" over the years. Today we find ourselves in the midst of a new turn that stresses relationality and multiplicity while also emphasizing novel political and ontological practices.

In this chapter we sympathize with this new academic orientation and with its emphasis on the relationship between people and food. We also believe that this contribution in agrofood studies has been intensely and strongly affected by the rise of "food movements," fueled in part by experiences that have the materiality of life as their basis. In other words, the interrelations between people and food also include the interconnectivity between popular perceptions of food and academic thinking. Nevertheless, we posit that this new current in agrofood studies has insufficiently involved Indigenous people in thinking about the affects and experiences with food in the contemporary world. We address this inadequacy by presenting an Indigenous perspective on these affects and experiences and how these in turn relate to food and the wider web of life. This is a perspective we think opens up unexplored conceptual terrain and allows imagining other possibilities.

Our empirical study in food takes place in the remote Colombian Amazon. There, the Andoque Indigenous people have experienced the introduction and proliferation of (illegal) gold mining, an activity which has had serious social and environmental impacts (including harmful effects on human health from exposure to mercury, changes in the landscape, distortion of the local economy, alcoholism, intra-household violence, among others) and which has elicited changes in their livelihoods. This chapter focuses on Andoque responses to the arrival of a new phenomenon in a historically plentiful Amazon: food shortage!

Central to our story is the disruption to the local "abundance," which the Andoque understand as what occurs when relations between all the entities (human and nonhuman) that dwell in their "world" or "territory" are balanced and reciprocal. As per this understanding, "abundance" maintains the flow of energy that is needed to sustain space. Importantly, the Andoque believe that food, as supplied by this territory, is "given" by spirit owners (of animals, plants but also of "inanimate" jungle beings such as rivers or rock formations). The Andoque seek reciprocal relationships with these spirit owners through the intermediation of *maloqueros* or shamans, who facilitate interactions between the multiple entities that supply abundance. When relationships between humans and nonhumans are balanced, spirit owners provide gifts that benefit the entire community, for example by offering plentiful animals to hunters, fish to fishermen, fruits to gatherers, or manioc to those opening up *chagras* (swiddens). Maloqueros and the Council of Elders reciprocate these gifts, for example by providing access to food (derived from swiddens and agroforestry) for all entities that comprise the territory according to the explicit behest of the spirit owners. In summary: when the spirit owners are happy that nothing is scarce, each body or entity be it individual, such as human, animal or plant, or a collective association, such as the territory, may satisfy its particular appetite and share abundance.

Recently, gold mining has changed the relations Andoque have with each other, with settlers or *colonos*, and with the nonhuman inhabitants of the forest and rivers (Verschoor and Torres, 2016). Indigenous leaders consider these changes as disruptive and a cause for a breakdown in the reciprocal relationships that give rise to abundance. To correct the situation, they have appealed to the spirit owners of the gold to hide the precious metal from miners. According to local tradition, this implies negotiating with these and other spirit owners about the way in which broken relations must be repaired and reassembled. In this chapter, we summarize a specific breach in the teleological notion of abundance. They describe this experience as a rupture that is associated with mining and associated negotiations that has led to food shortages. As a consequence, harm occurs when the "diplomacy" between entities is disrupted by mining activity changing the abundance into food scarcity.

We begin our account by first describing the study site and our methodological approach. This is followed by a description of local understandings and appreciations of abundance in a world of seemingly infinite resources that in recent time have become increasingly finite as a result of mining. We highlight a number of dilemmas that the Andoque face when choosing to take part (or not) in mining; choices that inevitably link up with their food supply. In the third part, we recount an emerging process of (re)negotiation in which Indigenous leaders try to restore abundance. We close the chapter with a discussion of some ontological elements that appear in our case, hoping to contribute to an understanding of the multiple ways in which "food" enters different worlds, an understanding that might enable ways of thinking about food and food futures that allow for a more generous and inclusive treatment of elements that are often eclipsed by the (perceived) necessity to give pride of place to reified notions of state, market or, for that matter, "the social."

Study site and methods

The Colombian Amazon occupies about 43 percent of the country (Gutiérrez *et al.*, 2004) and has high levels of ethno- and biodiversity. Our study site is located in the Middle Caquetá River basin of the Amazon and encompasses the Aduche Indigenous Reserve (*Resguardo*), home to a large number of the remaining members of the Andoque ethnic group (who call themselves "people of the ax" or */pʌʌsiʌ'hʌ/*). The Andoque live in a total of seven settlements, each characterized by large longhouses or *malocas* and surrounded by two to four smaller houses where individual families live. Formally these settlements are located within the *corregimiento* of Puerto Santander in the Amazonas Department (Echeverri, 2010 and 2015). Travelling to Aduche from Bogotá is done by air in a twin-engine aircraft and by outrigger boat or canoe, and may take several days depending on the availability of flights. Apart from a strong military presence around the airstrip, the presence of the Colombian government in the daily affairs of the Andoque and neighbouring ethnic groups is minimal to non-existent. Law and order in this vast and sparsely populated region is either in the hands of local institutions or the FARC-EP guerrilla.

Andoque political organization is consolidated through a Council of Elders, which is made up of the heads of five main clans: Hawk People (*kádannisiɵhɵ*), Sun People (*ixɵhɵ*), Ant People (*edaɵhɵ*), Deer People (*to'beósiɵhɵ*) and Beetle People (*xitasiɵhɵ*). The Council is responsible for protecting traditional ways of managing the territory and its members are mediators in the "diplomacy" that allows for abundant access to the goods offered by the spirit owners of different forest and river beings. Council members also have the task of determining how different territorial entities should coexist.

We carried out fieldwork during five trips of two weeks each to the Middle Caquetá between February 2012 and April 2016. We applied an ethnographic approach that was largely based on participant observation, followed up by interviews. Free, prior informed consent was obtained from all members (Indigenous and otherwise) who partook in the research.

The changing notion of abundance

The Andoque describe their territory, which they experience more as a body than as a cartographic or geographic entity, as the "land of plenty" (Griffiths, 1998). According to Don Mario, Chief of the Deer People and mediator between humans and spirits, their "body" is governed by a principle of relationality namely, the mutual dependence between all that exists (see also Londoño, 2012).

Andoque society is organized vertically and hierarchically. The vehicle for the reciprocal process of making offerings and engaging in spiritual transactions with the spirit owners is a ritualistic mastication of leaves from their most sacred plant: coca (*Erythroxylum coca*) (Farekatde, 2004). The Council also mediates processes of food production, and the members consider seasonality and personal competencies of farmers to be important elements when using the environment. According to the Andoque, whenever the existing state of reciprocity between beings (both human and nonhuman) is broken, an epidemic of diverse kinds of illnesses immediately breaks out, accidents or unexpected events suddenly occur, and shortages of game, fish, harvests, and forest fruit occur.

The Andoque's principal sources of food are swidden agriculture, fishing, hunting and, to a lesser extent, gathering. They create swidden farms or *chagras* (the fundamental family production unit) by clearing a small area in a forest, usually of one to two hectares, in the highlands or on a floodplain. The typical chagra cycle consists of a series of activities that may take place over a generation, such as clearing a forest patch, followed by a period of drying, burning the dry biomass (a first burning which takes about four hours in total followed by a second burning or *hoguereada* where the remains of the first burning are collected and burned), a recovery period of a few days to let the soil "cool," planting of up to 40 species of functional and utilitarian plants (but predominantly manioc and plantain), a growth period that includes between four and eight regular visits to eradicate weeds, harvesting (which is always done gradually except when large quantities of manioc are needed to make manioc beer offered at a *minga* or collective work party) and reseeding, especially of fruit trees, where productive units may exceed 40 years of intensive use (Peña-Venegas *et al.*, 2016). The Andoque complement farming with continual fishing and hunting for meat as well as gathering additional forest resources when needed.

The Andoque regulate the spaces within which its people may farm, fish, hunt and gather through zonification and rules of access, including forbidden, sacred and "collective use" sites. Forbidden areas are sites in which one cannot perform any activity; these are places inhabited by the creators and spirit owners of forest and river beings and are located in what the Andoque call "the Center" or place of origin, some *cananguchales* which are the sites of *Mauritia flexuosa* palm trees, and certain lakes, ponds, streams, hills, saltlicks, cemeteries and paths used by spirit owners. Sacred spaces consist of sites where one cannot enter without proper permission or offerings. These are spaces for ritual cleansing, purification, healing and harmonization, and include Devil's mountain, the nearby Araracuara rapids, and a number of cananguchales and saltlicks, paths, islands, and beaches distributed in the landscape. Collective use areas, in turn, are spaces where members of the community may establish chagras and pursue hunting, fishing, and recreation.

During a *mambe* consumption,[1] which characteristically lasted for much of the night, Don Mario explained the procedure for hunting animals in sacred areas:

> The animals take their food and water in places that man does not distinguish at first sight; one can only see them when one has taken mambe and that is why only men can go hunting. One can already see the place before entering because the animals demarcate their chagras [swiddens] with ropes. That's where they have their own coca, their manioc and their tobacco. One can also see the baskets in which they collect their food. The spirit owners of the animals have already spoken with us; it's them who show us the place … The place itself needs to be approached carefully; one first has to locate an animal's *maloca* which is always very close to their chagras; one should never look at the inside of the maloca but always at the outside, and one should only go as far as the door. It is at that moment that the animal offers itself to be hunted.

The intoxicating effect of the alkaloid amplified every detail of this narration; likewise one can intrinsically empathize with the concept of abundance that results from diplomacy exercised among different entities inside this territory. During the same mambe-taking session, Don Mario explained to one of his sons the basics of hunting, "Each animal is unique and has its own soul and spirit … That is why every animal must be respected. Everyone must show respect, which will guarantee that everyone has enough meat to eat."

Of course, the concept of abundance has undergone a continuous transformation. This has probably happened during moments that distort the "normal" course of events and that appear to repeat themselves cyclically. During such moments, different previously non-existing elements are incorporated into local mores and ways-of-doing, sometimes leading to an interruption in the reciprocal relations that supply and maintain abundance. For example, over the last 100 years, our study region has faced multiple bouts of resource extraction. These began in 1908 with the arrival of a rubber extraction company led by Peruvian merchants (Argüello, 1988; Landaburu, 2000; Echeverri, 2009) of which operations peaked in the early 1940s when demand for rubber skyrocketed during World War II (Uribe, 2013). The rubber boom was put to a halt for a number of decades after production zones shifted to Indochina and synthetic rubber was discovered. By the early 1970s the fur trade boom (Guyot, 1979) and drug production and trafficking (Gutiérrez *et al.*, 2004) entered the scene.

Mining for gold differs rather sharply from earlier resource booms. Starting in 2008, when entrepreneurs from outside the area introduced a number of wooden rafts and equipment to dredge the riverbed, it did not take long before its impact and effects were felt. The Andoque, especially men, are hired to work in shifts on the dredges. The activity has increased significantly. Early in 2012, some 14 rafts dredged for gold, and by the end of that year the number had increased to 34. In May 2015 some 150 rafts were said to "work" the Middle and Lower Caquetá River, a stretch of around 750 kilometers. At one point in that same month, we counted 25 dredges in a section that does not exceed two kilometers. At present, one can see a certain diversification in terms of raft sizes, extractive capacity, and division of labor. Local people are affected by the activities. One can see many men and women with respiratory and skin conditions (sores, rashes, and ulcers) that are clearly due to unprotected exposure to mercury fumes[2] that also ravage their nervous system.

Mining activities have also caused changes in the landscape through the dumping of surplus materials (especially mercury), destruction of riverine habitats, and environmental impacts due to the accumulation of heavy metals. Local people, however, tend not to talk about it and oftentimes cover their torsos or arms with t-shirts to hide the rashes and the ulcers. We overheard many conversations where people talked about medical complications that ended in women having spontaneous abortions. During conversations and informal chats, we also heard about violence and conflicts resulting from alcoholism, which is associated with the high consumption of liquor by miners who buy it with their wages. In this vein, an Indigenous woman, mother of three young children and daughter of Don Mario, stated the following:

> My husband went to work[3] more than two months ago and I don't know when he'll come back … I think he already has somebody else and certainly he does not care about the child we lost, nor the brother who drowned at work, and least of all about me … But I'm fine with this because when he would be drunk I just couldn't handle him, and that was every week … but yes, I miss him.

Such dilemmas appear and display coercive elements while at the same time generating dissonances that alter the ways of life of the Andoque. Linked to this and parallel to it, the concept of abundance changes and is transformed as new, foreign elements are integrated into the concept. Thus mercury, mobile phones, DVD players, outboard engines, imported clothing, alcohol, sports shoes, and so on are incorporated into a "new abundance."

For many, participating (or not) in mining activities proves to be a dilemma, as they have to choose between different, morally defined spaces in which different and often competing notions of abundance circulate. At present, many Andoque argue that it is not so bad to take a little gold to meet needs – an argument that is based on a perception of the infinite availability of resources. In these spaces, highly attractive incentives of the new abundance are put on display, weaving new interactions through all sorts of affects that seduce Indigenous hearts and minds. Yet when Andoque attune their ears to the "melodious songs of the sirens of gold," they face exclusion, punishment and rejection by the Council of Elders; their highest cultural authority. In this sense, Don Mario says:

> My five sons have worked. The two who love me are already with me and tell me that they can no longer work on that. The oldest left more than two months ago and I haven't heard from him since; he just went to work and left us and his wife and children … Another son doesn't talk to me and just

> wants the things one can buy with money from work … I do not want to talk to him either … And my youngest son drowned while working; he's buried right there – under the hammock you're now in.

Similarly, giving in to the seduction of the "accoutrements of modernity" may mean turning one's back on the gift of food that obtains from reciprocal relations with (the owners of) forest beings. Work on the dredges and the concurrent enjoyment of all manner of gadgets and beer means that chagras are left unattended and therefore deteriorate, and that less time or none at all is spent on fishing and hunting. Likewise, Andoque women have changed their lifeways and now leave their children to be taken care of by the elderly while they worry how to make ends meet and procure food. Thus we have observed the appearance of high consumption levels of substitute goods of a poor nutritional quality that were not heard of until recently (e.g., processed food, canned meat, and all manner of "artificial" nourishment). In a shop in Puerto Santander, the *corregimiento* closest to the Andoque communities, we counted over a period of 15 days a daily average consumption of 28 buckets of beer – which boils down to 672 beer cans per day. A few men together, not more than a dozen on any given day, thus spend around US$ 900 daily. This is quite an exorbitant figure given that in the jungle people do not normally have access to money.

From the perspective of Indigenous leaders, diplomacy and reciprocity with forest beings has been lost. As a result, their version of abundance has become distorted, walked away, mutated. The spirit owners of the forest and the rivers have been taken off-guard, bringing about an ontological collision between multiple realities (Latour, 2004, 2011, 2014). It is here, where both known and previously unknown scenarios meet face-to-face, where dilemmas that cause previously inconceivable dissonances appear. Again, as Don Mario explained, "That evil spirit of the Whites has stayed here to finish my people."

From Don Mario's perspective, which he shares with most elders, the obvious consequences of the deterioration of reciprocity between human and nonhuman entities are manifested through shortages of food, increased accidents in the forest, the loss of cultural identity, and a fragmentation of the territory.

Golden dilemmas

As we have shown above, some Andoque rebel against the new perception of abundance. Resistance was initiated by the Council of Elders the moment they saw that mining was affecting well-being in unwanted ways, most notably in the realm of food availability. "Where did things go wrong? What is the purpose of all of this?" elders asked. From their perspective, the abundance they were used to had been lost and food was no longer easy to obtain. In effect, the Andoque had moved from a world of infinite resources to one of food scarcity.

What can the Council of Elders, and those who subscribe to their point of view, do? What instruments do they have at their disposal to regain the "old abundance" so that food is plentiful once again? The answers to these questions are not clearcut and depend on the particular "reading" of the situation. In Andoque metaphysics, all denizens of the forests and the rivers (whether they be human or nonhuman, malignant or not) have a right to be there. Malignant beings, and Andoque know any number of them, can be of three sorts, and each one needs to be handled accordingly. We have malignant beings that only become so after a relationship goes sour. To redress the situation, this relationship should be renegotiated so that things go back to normal. This interpretation certainly exists: some members of the Council profess that gold has become evil because of unreciprocated greed. In this reading the problem can be repaired through different channels: joint excursions by the young and the elderly to sacred sites alongside the Caquetá River to see how the dredges continuously rape the riverbed so as to instill the idea of the "wrongs" that are carried out there, pointing to the lack of food whenever the opportunity arises, or constant admonishment by the elderly to those who carry out work, reminding them of their moral duty to re-establish a balance between the appetites of different beings.

Another type of malignant being, however, cannot be handled by humans: they are inherently bad and one cannot change their behavior. In these cases, it is better to avoid their dwelling places and prevent attacks.[4] Indeed, for some Andoque gold is an evil that was unleashed and that found a home in their territory. Running away from it would be the appropriate thing to do, but this is of no use as gold mining has become as ubiquitous as the forest. This is a totally new situation, and the Council has found a solution to it. As Don Mario told us, "We have initiated dialogue with the spirit owners of the gold so that they hide it in a place where the sunlight cannot touch it anymore … This will turn into a very long talk that will take a long time and a lot of mambe."

This line of repair is new and interesting but competes against other solutions that are linked to the third way a being can be malignant. According to the Andoque, there are forms of evil that are impossible to heal or to run or stay away from; rather, they are considered as a cancer that spreads relentlessly and knows no immediate cure other than a "total reset," meaning a cleansing of the world through cyclical events such as cataclysmic floods and droughts that clean what has been "fouled." Don Nelson expressed this as follows:

> At the beginning of Creation and before the origin of man there was this great flood, followed by a fire that was extinguished by another flood … This cleaned the world of all the bad things that were there before man appeared. This was the cleansing of the world … it happened on Crying Hill. After this, a hawk gave his first son his manioc and his coca … Nothing else was needed since this is all that is necessary to have all other things in the world.

For the Andoque, dealing with "the case of gold" is difficult and brings with it a number of dilemmas. Perhaps the most pervasive one relates to gold's contested status: is it a ticket to a new abundance that allows men to enjoy new things, or is it a threat to the old abundance and to the availability of food in particular? Another important dilemma (from the perspective of those who, like the Council of Elders, try to turn their back on the new abundance) is that of how to best "cure" the malady: what is the best way to do so when different, competing readings of evil exist? Is it a question of fine-tuning relationships between all entities involved in the gold complex? Or is it a question of asking the spirit owners of gold to take back their creature, much like the San in the film *The Gods Must Be Crazy* who go to the end of the world to throw away the contentious bottle of Coke that fell from heaven? Or should they wait until a catastrophic event ends it all?[5]

Discussion

We started our chapter by referring to Michael Carolan's assessment of the present state of agrofood studies. The field is obviously broad, but we agree with him that many recent studies on and about food, all with their particular inflections, highlight materiality, relationality, and multiplicity, while at the same time foregrounding political and ontological practices (see the introduction to this book for references). While we welcome these developments, it seems odd to us that so little attention has been paid to new ways of "thinking food." In this chapter we presented a case that may be "foreign" to many readers of food studies. We did this not to add exoticism to the field, but because we believe that the unexpected ways in which Andoque notions of abundance affect food in the context of mining can open food possibilities as well as serve as a source of inspiration.

We consider that the Andoque case allows for a number of conceptual considerations. At the cost of being selective, we point out three. First and perhaps most evidently, in agrofood studies the state and the market are given pride of place in explaining the ways people relate to food. The Andoque case, however, unfolds in a geographical space that is relatively untouched by the state or the market. We say "relatively," because it is evident that mining is the detonator of the loss of a certain form of abundance. Importantly, in our case both the market and the state play a negligible role in the protocol or governance mechanism to be followed to regain a specific form of abundance. While we are not suggesting that market and state do not have important influences elsewhere, the ways in which people relate to food work a bit like Occam's razor: it reduces all possible entities that may (unexpectedly) be attached to food to the relatively simple lexicon that the concepts of state or market allow for. To use a metaphor, it is as if one would filter gold grains through too coarse a sieve: the larger nuggets would no doubt be captured, but the smaller ones would fall through and be lost, together with the countless other particles to which they were attached. So perhaps, and this is an intuition, the twin concepts of market and state are too crude to account for the entities (the agencies or affects) that constitute webs of life?

Our second point is that, as we have shown, the Andoque food is embedded in a web of life and, because of its relationality, subject to politics. Yet, as our case also shows, for the Andoque the politics of food is not only a politics of *who*, but also a politics of *what* (Mol, 1999): a politics that is not only full of "traditional" human actors but also encompasses a host of nonhumans (animals, forest beings and spirits) and whose role is vital. As a result, the politics of life cannot be reduced to an anthropocentric, Foucauldian notion of biopolitics. Rather, it is a politics of life that involves multiple entities and realities and hence the exercise of Andoquian diplomacy – an ontological politics of affect and care or what Isabelle Stengers (2010) calls a "cosmopolitics." While these terminologies may appear foreign to the Andoque, it nevertheless appeals to the conceptual imagination to see that the Andoque have their own concept (abundance) with which to more or less make the same point: that matters of food are matters of concern that involve much more than can be captured by the notion of biopolitics.

In connection to this, and as a third consideration, we believe that the notion of abundance speaks to the conceptual imagination by the way it resonates with Actor-Network Theory and new materialism. Indeed, we believe that the Latourian notions of association, attachment, and proposition and the Deleuzian notions of affect, assemblage, or territory neatly overlap with the concept of abundance in that it completely dissolves boundaries between the natural and the cultural, between mind and matter, or between the far and the close. Just as in Deleuze's (1988) and DeLanda's (2002) flat ontology, the Andoque notion of abundance permits one to see how entities (in the Andoque case: that of a territory or body) change states and/or capacities as they become affected by the agencies of elements that were hitherto foreign to them, thus allowing for "lines of flight" or new abundances that extend the territory or body to other dimensions and allowing it to interact with other entities. These overlaps notwithstanding, it would be interesting to inquire about the extent to which Andoquian concepts (such as abundance) differ from Deleuzian or DeLandian ones. Our hunch is that, for the Andoque, abundance is not simply a concept that stands for, denotes or represents a world out there; rather, we understand it as a process: an empirical reality that needs to be *enacted,* to be *done* over and over again in order to become or remain real.

Yet, abundance is not only a (re)creation, but is also a catalyst for creativity. Abundance, in this latter sense, comes close to the Latourian notion of actor-network – which is "more like the name of a pencil or a brush than the name of a specific shape to be drawn or painted" (Latour, 2005: 143). Like a pencil that can be said to be "done" by atoms and molecules, Don Mario and the Council of Elders can be said to "do" abundance, and thus food, by passing on their knowledge (*el conocimiento propio*) and instructing young Andoques in the art of diplomacy. But a pencil also "does" (a drawing, a picture) in the same way abundance can be said to "do" never dreamt of, golden possibilities. And this might be an important message that could be put to good use in empirical food studies. We researchers *do* food by way of representation, but food itself (its vibrancy, its vitality, its affects) may possibly "do" a future politics of life outside a modern fix in which the only options seem to be a progressive but unsustainable "escape to the future" or a sustainable but archaic "return to the past." But what if there is no escape to the future because ecological mutations foreclose this possibility? And what if there is no return to a *Blut und Boden* past quite simply because this is a bygone era that has been obliterated by massive human destruction? What if, as Anna Lowenhaupt Tsing (2015) muses, the promise of cohabitation between humans and nonhumans lays not up or down the modern ladder (a future, a past) but rather sideways – a "Third Attractor" that Lovelock calls Gaia and the Andoque call Abundance? Perhaps it is time to entertain the idea that, rather than seeing the encounter with the Other as bumping into some relic from the past, we see it as a chance to learn? After all, the Andoque have been thrown into the modernist maelstrom and, *pace* Latour (2016), have difficulties that are just as grave as ours in orienting themselves within the notion of modernity. Maybe the Andoque can help us save ourselves?

Notes

1 *Mambe* is a powder made of dried coca leaves mixed with ashes of the *yarumo* tree (*Cecropia peltata*) which reduces the alkaloids and thus the narcotic effect of coca leaves. Use of *mambe* is exclusively for males and is the element that allows for establishing bridges between different species, the goal being the strengthening of territorial diplomacy. During the episode narrated above all men present (including one of the authors) were taking *mambe*.

2 Mercury is a key element to obtain gold; it is employed to assist with the extraction of gold from ore.

3 Expression used to refer to employment in artisanal mining activities.

4 In cases where these malignant beings choose a *maloca* or community to be their new dwelling place often the only way to deal with it is to relocate the community.

5 According to some Andoque who know full well how to articulate with Western discourses this is already taking place: loss of cultural identity, climate change, food scarcities, resource depletion, biodiversity loss.

References

Argüello, H. (1988). Prácticas agrícolas y consecuencias genéticas que permitieron una mejor adaptación de los indígenas de la Amazonía colombiana. *Agronomía Colombiana*, 5, 86–95.

Carolan, M. (2013). The wild side of agro-food studies: on co-experimentation, politics, change, and hope. *Sociologia Ruralis*, 53(4), 413–431.

De Landa, M. (2002), *Intensive Science and Virtual Philosophy*. London: Bloomsbury Academic.

Deleuze, G. (1988). *Spinoza: Practical Philosophy*. San Francisco: City Lights.

Echeverri, J. A. (2009). Siete fotografías: una mirada obtusa sobre la Casa Arana. In C. Chaparro and A. Chirif (eds), *Imaginario e imágenes de la época del caucho*, 42–57. Lima: CAAP.

Echeverri, J.A. (2010). Percepciones y efectos del cambio climático en grupos indígenas de la Amazonía colombiana. *Folia Amazónica*, 1(2), 85–93.

Echeverri, J.A. (2015). Substances and persons: on the culinary space of the people of the centre. *Tipití: Journal of the Society for the Anthropology of Lowland South America*, 13(2), 91–104.

Farekatde, N. (2004). *La cultura de tabaco y coca: análisis crítico sobre su reconstrucción socio-cultural, después de la explotación cauchera*. MSc thesis, FLACSO, Bogotá, Colombia.

Griffiths, T. (1998). *Ethnoeconomics and native Amazonian livelihoods*. PhD Thesis, University of Oxford.

Gutiérrez, F., Acosta, L.E., and Salazar, C.A. (2004). *Perfiles urbanos en la Amazonía colombiana*. Bogotá: SINCHI.

Guyot, M. (1979). La historia del mar de Danta, el Caquetá. Una fase de la evolución cultural en el noroeste amazónico. *Journal de la Société des Américanistes*, 66, 99–123.

Landaburu, J. (2000). Lengua andoque. In J. Landaburu (ed.), *Lenguas indígenas de Colombia: una visión descriptiva*, 275–288. Bogotá: Instituto Caro y Cuervo.

Latour, B. (2004). *Politics of Nature*. Cambridge, MA: Harvard University Press.

Latour, B. (2005), *Reassembling the Social: An Introduction to Actor-Network Theory*. Oxford: Oxford University Press.

Latour, B. (2011). Politics of nature: East and West perspectives. *Ethics & Global Politics*, 4(1), 1–10.

Latour, B. (2014) From ontology to deontology. *General Anthropology*, 21(1), 1–4.

Latour, B. (2016). *Reset Modernity!* Cambridge, MA: MIT Press.

Londoño, C.D. (2012). People of substance: an ethnography of morality in the Colombian Amazon? *Anthropological Horizons Series*. Toronto: University of Toronto Press.

Lowenhaupt Tsing, A. (2015). *The Mushroom at the End of the World: On the Possibility of Life in Capitalist Ruins*. Princeton, NJ: Princeton University Press.

Mol, A. (1999). Ontological politics: a word and some questions. In J. Law and J. Hassard (eds), *Actor-Network Theory and After*, 74–89. Oxford: Blackwell.

Peña-Venegas, C.P., Stomph, T., Verschoor, G., Echeverri, J.A., and Struik, P. (2016). Classification and use of natural and anthropogenic soils by indigenous communities of the upper Amazon region of Colombia. *Human Ecology*, 44(1), 1–15.

Stengers, I. (2010). *Cosmopolitics I*. Minneapolis, MN: University of Minnesota Press.

Uribe, T. (2013). Caucho, explotación y guerra: configuración de las fronteras nacionales y expoliación indígena en Amazonía. *Memoria y Sociedad*, 34 (17), 34–48.

Verschoor, G. and Torres, C. (2016). Mundos equivocados: cuando la "abundancia" y la "carencia" se encuentran en la Amazonía colombiana. *Íconos – Revista de Ciencias Sociales*, 20(54), 71–86.

4

ENCOUNTERS WITH THE BRAZILIAN SOYBEAN BOOM

Transnational farmers and the *Cerrado*

Andrew Ofstehage

Introduction[1]

The production of soybean in South America covers 57 million hectares (Oliveira and Hecht, 2016). For critics it is an exemplar of the corporate food regime (McMichael, 2010) and the great transformation of rural South America (Gudynas, 2008), while its proponents call large-scale soya a model of how to feed a growing world population (Economist, 2010). Often lost in abstract narratives of the soy boom arc, however, is the situated reality of soy growers and their production. Reflections on the neo-extractivist governments of South America and transnational corporations place soy production within a set of social, economic, and cultural transformations, but by focusing on capital and mechanisms of capital, they mask the improvisation and creativity that contest and enable flows of capital, thus implicitly accepting capital as the source of life, rather than a parasite of life (Braun, 2015; Paredes, Sherwood, and Arce, 2015).

The situation of industrial soy farming in Brazil is a seemingly straightforward assemblage. The nation-state supports soy production through favorable public policy. Soy farmers seek out profit, and social and environmental activists organize to block soy expansion. Soy producers implement the "Brazil model" of farming, share a common vision of profit and growth, and act as economically logical characters. A cosmopolitical perspective of soy in Brazil is more complicated and less capital-centric (introduction to this volume). Abstractions of the Brazil model and *Soylandia* fall apart at the level of daily practice and contingencies of soy production emerge. Farming practice depends as much on the soil quality as the disagreement between an agronomist and labor foreman; farmer decisions are made to maximize profit but also to impress neighbors; soy farmers spread socio-environmental harm, but also flee crisis. The migration of U.S. farmers to the fast-growing Mapitoba[2] region of the Brazilian *Cerrado* is one case of the contingent, relational, and everyday unfolding of the soy boom. Their migration to Brazil and work within the soy boom straddle two distinct macro-economic processes – the soy boom in Brazil and the farm crisis in the United States, but are also products of their own agency, plans, and improvisations that give life to these processes.

In this chapter, I draw on "deep empiricism" (Stenner, 2008; Whitehead, 1978) by way of ethnographic detail to analyze abstractions of soy farming. Reification of the soy boom as a bounded "thing" limits deep engagement with a complex and messy entanglement of actors, values, and practices. Beginning with the lived realities of transnational soy farmers' migration, I arrive at the expansion of soy frontiers in the Brazilian Cerrado, personal narratives of crisis, social values of work and farming, and the material realities of soy production. I focus on a group of migrant farmers from North America involved in the ongoing processes of crisis and boom, with the questions: how does one become a transnational farmer and what transformations and continuities does this beget? Early in my studies I expected to find one of two realities. First, American farmers re-created the Midwestern farm with its grain silos, red barns, homestead, farming practices, and ideals of "good farming." Or second, American farmers adopted the Brazil model and all that came with it – farm management practices, living in urban centers, and *churrascarias* (a Portuguese "barbecue" cook-out) on Sunday evenings. I found neither of these extremes, nor did I find something between the two. Through ethnography, I came to see the transnationalization of farming as the creation of entirely new subjects, rather than the simple adoption or continuation of farming practices and living styles. Thus this chapter asks: what subjects emerge from the transnationalization of farming?

Methods

Soy production in South America has brought massive social, economic, and environmental transformations. Between 1986 and 2002 farmers transformed a third of native Brazilian Cerrado into agricultural land (Brannstrom *et al.*, 2008), and soy production in the Amazon has contributed to deforestation as well as financed an expansion of infrastructure (roads, power, railways), facilitating the development and the rise of other extractive industries in the biosphere (Fearnside, 2005). The transformation of the Cerrado from savannah to farmland is a product of stricter enforcement of protections against rural development in Amazônia and incentivization for farming in the Cerrado (Oliveira, 2013).

In order to understand the on-the-ground experiences of this transformation, I turn to the interests, values, and agency of actors involved in soy production in order to make visible the interfaces or the spaces where actors encounter other lifeworlds (Arce and Long, 2000). This actor-oriented view of agrarian change re-orients the classic economic hierarchy of states and markets as powerful and actors as weak, by placing them on the same plane of action and re-conceptualizing their relationship as co-creative (at times cooperative and at others conflict-ridden), rather than one of outright dominance by the structural forces of political economy. Agrarian change rarely goes according to plan. Rather, it is filtered through actor-interfaces and contestations of knowledge and practice. Likewise soy production in South America is not only a product of global market and nation-states, but also of the ways that farm families, indigenous groups, and other collectivities actively engage with these influential actors. Via deep empiricism I construct a narrative from these actor engagements with soy production and contribute to recent studies on the heterogeneity of farming styles, business organization, and social values among Brazilian farmers and ranchers (Hoelle, 2015; Mier and Terán Giménez Cacho, 2016; Ofstehage, 2016; Wesz Junior, 2014). Heterogeneous farming organization and styles within a similar political-economic sphere of soy production express different values and goals as well as different material realities of production.

My dissertation research is centered on the experience of transnational soy production, which led me to ask: do transnational soy farmers participate in community making in Brazil? Do they forsake U.S. based communities? How do they reconcile American and Brazilian modes of production and values of work? How do farmers experience the everyday reality of transnational agriculture? To understand family life, I placed as much importance on attending birthday parties and Thanksgiving festivities as tillage practice. In focusing on everyday life I re-centered my study from the level of Soylandia to the level of situated action.

Research was conducted over two research periods – May to June 2012 in the U.S. Midwest and Bahia, Brazil and September 2014 to December 2015 in Goiás and Bahia, Brazil. During this time I worked with a group of North American family farmers, predominantly from the U.S. Midwest, who had migrated to the Mapitoba region primarily in the 1980s and early 2000s. I conducted semi-structured interviews with family members, participant observation during farm and office work, and online archival research on farmers' blogs, websites, and newspaper articles. For background material, I took note of relevant public policies that encouraged soy production, tracking flows of capital in and out of Brazil, and the expansion of American-owned farmland in the Cerrado.

Transformation in the *Cerrado*

President Getulio Vargas' "March to the West" created the concrete airplane-shaped capital of Brasilia. It also created soy fields in the Cerrado – a tropical wooded savannah in central Brazil. Similar to processes of populating and claiming Brazilian Amazônia and laying claim to "unproductive" Amazonia in Bolivia, public policies in the Cerrado incentivize extensive soy production in order to populate and make productive territory where previously there was "nothing," according to many U.S. and Brazilian farmers in the region. Amazônia and the Cerrado to its south have a dynamic relationship of taking and retaking land at their border as climate and weather patterns change. Their intertwined ecological history is matched by their similar soy production histories.

Incentivization of Cerrado production stands in distinction from soy production in Amazônia. Successful activism of Greenpeace and other NGOs has induced soybean importers to demand proof that soy does not originate in Amazônia. Strict government protection, industry watchdogs such as the Roundtable on Responsible Soy, and non-governmental organizations keep a vigilant eye on Amazonian production. Meanwhile, the Brazilian state has encouraged soy production in the Cerrado. Legal land reserves in Brazil require landowners to set aside land. In Cerrado states, farmers are required to set aside 20–35 percent of their land, depending on the state, while farmers within the Amazonian ecosystem are required to set aside 80 percent.[3] Infrastructural improvements and government-funded agricultural research further induce farmers to make the Cerrado "productive" (Nehring, 2016; Oliveira, 2013). The incentivization of the Cerrado to soy farmland transformation, via legal land reserves and government incentives for Cerrado soy production has also pushed rural Bahian communities from *gerais* (the commons) on which they depend for subsistence agriculture and ranching.

Brazilian government research into suitable seed varieties and soil management processes made soy production in the Cerrado possible and profitable, while government support programs induced farmers, primarily from the south of Brazil, to migrate to the northeast of Brazil. The late 1960s military government enacted an agrarian development project, POLOCENTRO, to incorporate the Cerrado into the national economy and support rural producers through funding for infrastructure development and agricultural research. Concurrently, the government provided loans for land purchases on generous terms through the PROTERRA program (Jepson, Brannstrom, and Filippi, 2010). U.S. soy farmers, as well as gauchos from southern Brazil, found large contiguous tracts of land, low land prices, strong infrastructure, cheap labor, and reduced presence of environmentalism compared to Amazônia (Mondardo, 2010). Western Bahian agricultural production increased from 2.05 km^2 in 1979 to 1,615 km^2 in 1986 then to 5,743 km^2 in 2000 and to 7,259 km^2 in 2005. By 2005, agricultural production accounted for 55 percent of total land (Brannstrom, 2009). Meanwhile, the introduction of capital and self-identified "agents of modernization" provoked a transformation towards agricultural capitalization, *tecnification*, and in farmers' words, modernization (Diniz, 1984).

A many faced crisis

With headlines like "American farmers try their luck in Brazil" (Omestad, 2008), "U.S. farmers scramble to buy Brazil's farmland" (Elizondo, 2012), and "U.S. farmers put roots down in Brazilian soil" (Romero, 2002), the arrival of U.S. farmers in Brazil has been greeted with a blend of pity and curiosity. The headlines provide an insight into this migration – the search for opportunity and a second chance in Brazil. The articles' texts, like the excerpt below, allude to histories of settling frontiers, adventure, and hopes of profit:

> For a few decades, Brazil has been home to a few communities of American Mennonite farmers. But a newer wave of mostly young farmer-investors has resolved to head south for profit – and, perhaps, for the challenge of facing down obstacles more reminiscent of the American frontier of the past than of the U.S. farm belt of today. "What we're doing here isn't all that different from what my great, great-grandfather did," says Matthew Kruse, a native of Royal, Iowa.
>
> He now directs operations on three farms totaling 23,000 acres for an outfit called Brazil Iowa Farms. He first visited Brazil in 2001, then moved here permanently in 2004. He has a Brazilian girlfriend, and out on the farms he manages between 90 and 150 workers, depending on the time of year. When he first visited rural Brazil, he says, "It seemed like we were looking at something in National Geographic. We had fallen off the planet. Now, I call it home."
>
> *(Omestad, 2008)*

What contradicts this view of farmers as frontier settlers and adventurers are the dual realities of crisis in the United States and risk-free opportunity in Brazil. Farm journal magazines in the late 1990s and early 2000s presented Brazil as the new center of soy production and a land of unrivaled productivity, beneficial climate, and inexpensive labor. The only drawback was poor infrastructure. Further, for politicians in Brasilia and both local and foreign agro-industrialists, the Cerrado of Brazil represented a wasteland waiting to be developed: land that could be purchased in massive tracts for pennies. The common narrative in farm journals, and often expressed by transnational farmers still, was the inevitability of Brazil overtaking the U.S. in soy production once the Cerrado reached its full potential. Alongside narratives of farm crisis in the United States, Brazil became an ideal site to reproduce not only farming histories, but histories of migration and settling barren wastelands. The imaginary world of the Cerrado, for migrant farmers, was a wild, untamed, and neglected space ready to be made productive. It represented an opportunity to continue farming, migration, and pioneer histories as well as a chance to turn a profit (Ofstehage, in press).

In the United States, farmers faced high farmland value, farm consolidation, which left little available land, and expensive startup costs. Average farmland prices have doubled from U.S.$ 1,000 per acre to U.S.$ 2,000 per acre since the 1980s (Nickerson *et al.*, 2012). Land values are even higher on fertile, productive land. High quality Indiana farmland, for example, exceeded U.S.$ 9,000 per acre in 2013 (Dobbins and Cook, 2013). An Idaho farmer, who by the time of our interview had lost his Brazilian farm explained, "We had the choice between going broke in Idaho or going broke in Brazil." Another family sold farmland in Northwest Iowa in 2001 for U.S.$ 3100 per acre and bought farmland of similar quality for U.S.$ 800 per acre in Western Bahia. With the rise in farmland prices farm size has also increased. In 1982 the average U.S. farm size was 589 acres; in 2007 it was 1,105 acres. The bloating of the U.S. farm can be credited to labor-saving technologies such as GMO crops, reduced tillage, and equipment that allow farmers to manage larger acreages. Farms have also shifted towards specialization in which they focused on the production of fewer commodities (MacDonald, Korb, and Hoppe, 2013). Accompanying this shift had been an increasing corporatization of U.S. farms, as family farms become incorporated as limited liability corporations or publicly traded corporations (O'Donoghue *et al.*, 2011). Together, farmland value increases and high rates of farm consolidation reduced land access, especially for young farmers.

When I asked a farmer why he migrated to Brazil to farm when he could have found non-farm work, he paused before saying, "I don't know." Another long pause, then "I've never, I grew up farming and … I never had any doubt I was gonna farm." An interest in continuing family farming legacies was present in many farmers' narratives of deciding to migrate. Often without prompting, farmers would make their own connection between this process and their family migrations from Europe to the U.S. Midwest. This landscape of crisis interconnects with the political economy of Brazil to co-construct the migration of North American soy producers to Bahia and the emerging soy frontier of Mapitoba. Faced with rising land prices among other problems, young U.S. farmers, looking for a piece of land to farm, had turned to Brazil where they find prices for as little as a sixth of land prices in their home counties. Looking into specific career histories we could see the interplay between U.S. crisis, farming values and opportunity in Brazil.

Dennis was a fourth generation farmer in Illinois and began farming on his own in 1979. He "worked his butt off to make fifty dollars an acre" on corn and soybean farming 160 acres, and since he quit farming in Illinois in 2000, farming had gotten even more competitive and profits were much lower. In the 2000s, he began to transition from farming towards real estate, specifically farmland sales. He visited Brazil on an Illinois farmer leadership program in 1998, then again in 2003. He noted that Brazil was expanding rapidly and consultants were saying that in a year or two they would out-produce the United States, although he later complained, "they still say this." While fact finding he met a couple in Tocantins, Brazil that were interested in selling. He bought and opened a test plot soon after, in 2004. Tocantins also drew his interest for its relative lack of American farmers compared to Bahia. He found the culture of the U.S. farming community of Bahia distasteful. The test plot yielded well enough to encourage him to purchase a 1,010 hectare farm the next year, with financial support from a small group of investors. As of 2012, Dennis farmed 650 hectares, of which nearly 550 hectares were planted to soybeans. He rented 700 hectares, of which 660 hectares were planted to soybeans. Dennis and his agronomist also managed 280 hectares and he and his investors had plans to add 1,800 hectares to their operation, though since this time they had decided to sell the farm. For Dennis, Brazil offered an opportunity to make economic profit, but also become involved again in farming. In the United States, farms of 100–249 acres, including Dennis' original 160 acre farm, made an average profit of 1.2 percent on corn production and lost an average of 0.03 percent on soy in 2013, far behind larger farms (MacDonald *et al.*, 2013). However cheaper land and labor prices as well as access to capital allowed him to profitably farm in Tocantins.

Leon owned a small farm in Idaho until realizing that he could not make a living off of his 600 acre edible bean and sugar beet farm in the United States and wanted to build something for his children. After four consecutive years of financial loss on the Idaho farm, Leon went on an agricultural tour of Bahia, Brazil, found impressive farming opportunities and farming scale, sold his home acreage, and purchased a 2,300 acre farm near Luis Eduardo Magalhães in 2007. Less cautiously and slowly than Dennis, Leon began farming soon after arriving in Brazil, though he later regretted his fast pace of farm purchase, clearing, and planting. Another farmer blamed Leon's difficulties on intentional obstruction by another U.S. farmer in the area, who dragged his feet in delivering purchased machinery. Disputes of this nature are not uncommon. Difficulties with paperwork, securing farm machinery, and weather delayed his first year's planting beyond an acceptable growing year length and soon after starting the farm Leon went bankrupt and had to return to the United States. Though he does not own a farm today, he hopes to either purchase a farm or engage in farming as a farm worker in order to return to agricultural work. Leon, more so even than Dennis, was driven by an interest in the farming livelihood. His choice when facing bankruptcy in Idaho was not between going bankrupt or finding non-farm work; it was between going bankrupt farming in Idaho and possibly going bankrupt farming in Bahia. As he recounted his early visits to Brazil, he was not impressed by the rate of profit or economic risk, but rather the scale and the possibilities available to him to continue farming.

In each career history interview, transnational soy farmers mentioned the lack of opportunity for young farmers in the U.S. Midwest and the wealth of possibilities in Brazil. Farm journal magazines contribute to this perception in their portrayal of Brazil as a land with a warm climate and good rainfall, where land, labor, and capital are cheap, and the only limitation was infrastructure. Farmers recounted first Brazilian experiences, often on organized farm tours, with a sense of awe and wonder that was rarely present when talking about the realities of farming before or after their migration. According to a tour operator who specialized in leading Midwestern farmers on agro-tours of Argentina and Brazil, many farmers talked about the sight of felling trees as their favorite aspect of the trip. Others talked about the result of that felling, the sight of flat land as far as they could see, as the most awe-inspiring.

These early tour experiences reinforced the imagined world of the Cerrado as a wasteland landscape with endless potential for development and productivity. The sense of hope in continuing farming, albeit on somewhat different terms, despite the crisis as well as the sense of participating in the "development" of the Cerrado from "nothing" to something, is still palpable on my interview recordings. Nevertheless, the terms of their farming did indeed change. As described by Gudynas (Gudynas, 2008), the production of soy, incentivized by neo-extractivist public policies, had ignited a shift from traditional farming towards farming without farmers. This transformation entails the decoupling of the farmer from farm work, land, and exchange and a shift towards *management* of labor, renting of land, and full commodification of social relationships, production, and labor. In a sense, this was not a break with the farming experience of North Americans, but a continuation or perhaps acceleration of an ongoing process.

While the farm crisis of the 1980s led farmers to foreclose and lose their farms, those who continued to farm saw a transformation of practices. Reflecting Berry's (1977: 45) cultural crisis of farming as a "revolution in the farmers' mind" and transformation from land husbandry towards finance and technology, farmers shifted away from "well-tended fields, hard work, and performing tasks 'in a timely manner'" as farm tasks and skills worthy of praise (Barlett, 1993: 238). This also preceded Gudynas' "great transformation" in South America (Gudynas, 2008). The introduction and adoption of biotechnology, reduced tillage, and large and technology-advanced (even GPS-guided) machinery, Midwestern farmers often claimed, had made "good farmers lazy and lazy farmers good." Since 1982 operator labor had fallen by 40 percent and hired labor had fallen by 30 percent (MacDonald *et al.*, 2013). Operator and spouse labor accounts for three-fifths of labor on small farms, one-half on medium-sized farms, and 11 percent on large farms – meaning that large farms used an average of 89 percent hired labor and even small farms hired out two-fifths of their labor (MacDonald *et al.*, 2013). Many, 40 percent, work off-farm (MacDonald *et al.*, 2013), some depending on this work for financial survival. Processes of financial change had narrowed profit margins, increased dependency on finance markets, shifted expertise from agro-ecological knowledge to knowledge of high finance, and reduced labor requirements (MacDonald *et al.*, 2013). This process was only accelerated for many U.S. farmers in Brazil as their working conditions, materials, and limitations had produced a new set of valued skills, knowledge, and values, emerging together as a new taskscape for migrant soy farmers (Ingold, 2011).

When I asked Dennis about his greatest challenges of farming in Brazil, he did not mention the increased pest pressure, the iron-heavy latosol soils, or the different climate. Neither did he complain about cultural barriers such as language, social relations, and customs. His greatest challenge was labor and environmental regulations. Employing 20 workers, one of his main worries was complying with strict labor regulations that govern working hours, vacation days, worker termination protocol, and living conditions for workers living on the farm. In addition to this, environmental regulations required him to set aside around 50 percent of his land due to minimum set-aside laws, water source protections, and protections for hardwood trees. While recognizing the need for regulations to protect vulnerable and valuable ecosystems such as the Amazon, he dismissed environmental regulation of the *Cerrado* as governmental overreach and restrictive of growth.

For agronomic decisions, Dennis turned to his full-time Brazilian agronomist for consultation and for the agricultural labor he turns to his 20 workers. Dennis was able to reside primarily in Illinois, traveling three to four times per year to Brazil, and making management decisions via Skype. The taskscape of Dennis' work had become one based in an office from which he could email or Skype investors, farm managers, and agronomists. It was a mobile space as well as he frequently travels to Tocantins and back to check on farm work progress. His tasks, reflecting Gudynas' great transformation, were those of a manager – directing farm labor rather than doing it and controlling land rather than tilling it. As farm manager he was unbound from the farm, the land, and the labor, yet he maintained an identity as a farmer and as part of a farming tradition. In short, Dennis had embraced the taskscape of Brazilian farming. He moved his work site from the field or tractor to his personal office (which happens to be 6,000 kilometers away from his fields) and shifted daily tasks from direct farm labor to managerial oversight while working in a different biosphere and political economy.

Leon, on the other hand, attempted to employ a more "American style" production strategy. This entailed living on his farm, rather than in the nearby frontier town Luis Eduardo Magalhães, doing his own farm work, making farming decisions as a family, owning all of his production land, and self-financing. While this strategy allowed Leon to engage with agriculture in the way he preferred – he valued farm work over managerial office work – it also led to ridicule from fellow farmers, both Brazilian and U.S.-based, for his provincial attitude towards farming. Speaking with him several years after he returned with his family to the United States, Leon spoke passionately about his emotional connection to farming and to Brazil. For him, it is not enough to be involved in farming from a distance. He wanted a connection to land and production via his own labor and to have land to pass on to his children to continue family farming traditions. The slow pace of business, particularly dealing with paperwork for land and equipment sales, and his undercapitalized farming operation made farming difficult for Leon.

Unmentioned, though also playing a part, was the disconnect between how farming in Brazil is portrayed on farm tours and farm journals and the actual day-to-day challenges of farming. Leon needed to reconcile his knowledge and skills of farming acquired in Idaho with the on-the-ground realities of farming in a different climate, with different crop pests, and wholly foreign soil profiles which required different soil amendments. Leon's misunderstanding of appropriate planting time frames contributed to the eventual demise of his Brazilian farm. The taskscape of Brazil, unfortunately for him, involved an adaptation to bureaucratic practices as well as agronomic practices and he was unable to adapt to either.

As noted elsewhere (Ofstehage, 2016), farming in Brazil was relatively easy; it's becoming Brazilian that's hard. For most U.S. farmers in Brazil, a team of agronomists and managers aided decisions on agronomy and local laborers carried out the work. Farming was rarely listed as a challenge for these farmers. Becoming a manager of workers, however, and all the tasks that entailed, was another story. Along with adapting to the material realities of farming in the Cerrado by reducing tillage, increasing pesticide use, and adopting new seed varieties, many also adopted cultural aspects of the Brazilian model of industrial agriculture. The families I worked with often listed government labor and environmental regulations as their primary challenge, and managing workers was a close second. Many spend the majority of their working time in an office, and when they are out of the office, it was primarily to oversee work being done. The crisis of farming, then, was hardly an economic crisis alone, but rather to paraphrase Jason Moore, it was a socio-ecological-cultural crisis.

Conclusion

The story of U.S. farmers producing soybeans in the Brazilian *Cerrado* begins not just in the green corn fields of Illinois, the rolling potato fields of Idaho, or the flat savannah of Bahia. Likewise, the story is neither driven by capital, nor by its absence. Rather, this story is largely one of contingency, improvisation, and unpredictability. The farmers reached a crossroads in the U.S. farming sector at which they could leave farming for wage labor, incur large debts for farm expansion, or transnationalize. In choosing the latter, they opened up a new arena for negotiation of farming values, practices, and community making, effectively redefining farming and what it is to be a farmer. This everyday negotiation of *Soylandia* is indicative of the hidden realities of development that can be at least partially revealed via deep empiricism and attention to the on-the-ground experience of rural development in South America. The topic of large-scale soy farming in the neglected but vulnerable Cerrado is in many ways an unpleasant one, but one which helps to understand food as it is, and not, as the editors of this book argue, as one would like it to be.

The story of transnational family farmers in Brazil is a paradox. Capital has enabled families to expand their soy production and contribute to the growth of the soy industry as well as produce associated ecological, social, and economic consequences. At the same time, more than capital is behind this activity. Both Leon and Dennis, farmers whose stories reflect the migrant U.S. farmer population in general, were motivated to migrate as much by the possibility of wealth accumulation as they were by the desire to continue farming as a vocation in the face of crisis. Socio-ecological crises emerge where fiscal, agro-ecological, and political crises thwart strategies for capital accumulation and render technologies, labor, and techniques of capital accumulation inadequate. Social and environmental crises are constitutive of commodity frontier expansion just as agrarian crisis begets destruction, but also opportunity. It is through frontier expansion that capitalism thrives on crisis (Moore, 2010).

This observation compels me to return to Braun's argument at the beginning of this chapter: capital is not determinate, but rather "parasitic" to the life of farmers. Capitalist expansion in the Cerrado is not occurring due to capital alone, but capital in conjunction with the ever-emerging values, skills, and knowledge of rural families desperate to re-emerge from a crisis of farming. Paradoxically, or perhaps predictably, the migrant farmer's space to maneuver within the farming crisis has enabled the vibrant expansion of soy production. Dennis responded to declining farm profits in Illinois by purchasing Cerrado land in Tocantins and redefining the nature of his work. He escaped his own economic crisis and at the same time further entrenched soy production at the margins of Soylandia. Leon recalled a difficult choice between going bankrupt farming in Idaho and going bankrupt in Bahia. A farm crisis in the United States gave impetus to his migration to Brazil, imaginations of a farming utopia in the Brazilian Cerrado and a deep connection to farming as a way of life.

Deep empiricism (Stenner, 2008; Whitehead, 1978) of the migration of North American soy farmers to the Cerrado enables a deeper understanding of the soy boom and its characters via grounded experiences and de-centers capital as the primary driver of agrarian change and action. Dennis and Leon's migration to Tocantins and Bahia is not a seamless transition and transformation, but a tricky interplay of rural social life. Again, this interplay results in neither a wholesale adoption of Brazilian farming practices, values, and styles, nor a simple translation of U.S. farming lifestyles, nor even a hybrid of the two. New subjectivities of the transnational farmer and the wasteland-turned-productive Cerrado emerge out of this interplay as farmers and land react against each other. A farmer once dismissed my question on his forest reserve, a piece of the Cerrado owned by the farmer and set aside according to environmental regulations, saying, "I've never even seen it." The same farmer claimed some pride in having made the Cerrado productive. Idiosyncratic relationships between farm and land, both antagonistic and heroic, highlight the messiness of the soy boom and the emergence of new subjectivities.

Critical food studies enable a re-opening of dialogue around the multiple possibilities and consequences of rural development. Even greater, they enable an entirely new set of questions around the emergence and constitution of rural development. The migration of U.S. farmers to Brazil is indicative of the ever-emerging reality of Soylandia – new sets of values and practices arise out of new engagements between farmers, land, politics, and capital. This critical study of transnational soy farmers challenges scholars, activists, and students to consider the vibrant social worlds of even seemingly fully capitalized industrial farming. Industrial soy farming in Brazil is too easily dismissed as a battle between destructive capital and vulnerable rural communities and fragile ecosystems, but beginning from empirical data rather than reified categories of bad farmers and good victims, we can better understand the tricky interplay of actors, land, and capital that co-create this formulation of being a farmer and farming in Brazil.

Notes

1 The author thanks Fulbright-IIE, the Wenner-Gren Foundation, and the Graduate School at the University of North Carolina at Chapel Hill for the funding that supported this research.
2 Made of up the states of Maranhão (MA), Piauí (PI), Tocantins (TO), and Bahia (BA).
3 For example, a Tocantins farmer who purchases 1,000 hectares of native Cerrado land must keep 350 hectares out of production, plus any land that is adjacent to open water. A Bahian farmer would have to keep 200 hectares out of production, plus any land adjacent to open water.

References

Arce, A. and Long, N. (2000). Reconfiguring modernity and development from an anthropological perspective. In Arce, A. and Long, N. (eds), *Anthropology, Development and Modernity*. London: Routledge, 1–36.

Barlett, P. F. (1993). *American Dreams, Rural Realities: Family Farms in Crisis*. Chapel Hill, NC: University of North Carolina Press.

Berry, W. (1977). *The Unsettling of America: Culture and Agriculture*. San Francisco, CA: Sierra Club Books.

Brannstrom, C. (2009). South America's neoliberal agricultural frontiers: places of environmental sacrifice or conservation opportunity. *AMBIO: A Journal of the Human Environment*, 38(3), 141–149.

Brannstrom, C., Jepson, W., Filippi, A. M., Redo, D., Xu, Z. and Ganesh, S. (2008). Land change in the Brazilian Savanna (Cerrado), 1986–2002: comparative analysis and implications for land-use policy. *Land Use Policy*, 25(4), 579–595.

Braun, B. (2015). The 2013 antipode RGS-IBG lecture: new materialisms and neoliberal natures. *Antipode*, 47(1), 1–14.

Diniz, J. A. F. (1984). Modernização e conflicto na fronteira ocidental do Nordeste. *Revista GeoNordeste*, 1(1), 12–20.

Dobbins, C. and Cook, K. (2013). *Purdue Agricultural Economics Report* (August 2013). West Lafayette, IN: Purdue University. Retrieved from www.agecon.purdue.edu/extension/pubs/paer/pdf/PAER8_2013.pdf

Economist. (2010, 26 August). The miracle of the Cerrado. *The Economist*.

Elizondo, G. (2012, 29 September). US farmers scramble to buy Brazil's farmland. *Al Jazeera*. Retrieved from www.aljazeera.com/indepth/features/2012/09/2012913112137744956.html

Fearnside, P. M. (2005). Deforestation in Brazilian Amazônia: history, rates, and consequences. *Conservation Biology*, 19(3), 680–688.

Gudynas, E. (2008). The new bonfire of vanities: soybean cultivation and globalization in South America. *Development*, 51(4), 512–518.

Hoelle, J. (2015). *Rainforest Cowboys: The Rise of Ranching and Cattle Culture in Western Amazônia*. Austin, TX: University of Texas Press.

Ingold, T. (2011). *Being Alive: Essays on Movement, Knowledge and Description*. Abingdon: Taylor and Francis.

Jepson, W., Brannstrom, C. and Filippi, A. (2010). Access regimes and regional land change in the Brazilian Cerrado, 1972–2002. *Annals of the Association of American Geographers*, 100(1), 87–111.

MacDonald, J. M., Korb, P. and Hoppe, R. A. (2013). *Economic Research Report Number 152: Farm Size and the Organization of U.S. Crop Farming* (Economic Research Report No. 152). USDA Economic Research Service. Retrieved from www.ers.usda.gov/publications/err-economic-research-report/err152/report-summary.aspx#.UjSslWR4bfU

McMichael, P. (2010). Agrofuels in the food regime. *Journal of Peasant Studies*, 37(4), 609–629.

Mier y Terán Giménez Cacho, M. (2016). Soybean agri-food systems dynamics and the diversity of farming styles on the agricultural frontier in Mato Grosso, Brazil. *The Journal of Peasant Studies*, 43(2), 419–441.

Mondardo, M. L. (2010). A "territorialização" do agronegócio globalizado em Barreiras – BA: migração sulista, reestruturação produtiva e contradições sócio-territoriais. *Revista NERA*, 13(17), 112–130.

Moore, J. (2010). "Amsterdam is standing on Norway," Part II: The Global North Atlantic in the ecological revolution of the long seventeenth century. *Journal of Agrarian Change*, 10(2), 188–227.

Nehring, R. (2016). Yields of dreams: marching west and the politics of scientific knowledge in the Brazilian agricultural research corporation (Embrapa). Retrieved from www.iss.nl/fileadmin/ASSETS/iss/Research_and_projects/Research_networks/ICAS/35-ICAS_CP_Nehring.pdf

Nickerson, C. J., Morehart, M., Kuethe, T., Beckman, J., Ifft, J. and Williams, R. (2012). *Trends in US Farmland Values and Ownership*. USDA Economic Research Service. Retrieved from www.ers.usda.gov/media/377487/eib92_2_.pdf

O'Donoghue, E. J., Hoppe, R. A., Banker, D. E., Ebel, R., Fuglie, K., Korb, P., Livingston, M., Nickerson, C. and Sandretto C. (2011). *Economic Information Bulletin Number 88: The Changing Organization of US Farming*. Washington, DC: USDA Economic Research Service.

Ofstehage, A. (2016). Farming is easy, becoming Brazilian is hard: North American soy farmers' social values of production, work and land in Soylandia. *The Journal of Peasant Studies*, 43(2), 442–460.

Ofstehage, A. (in press). Agricultural financialization and the 21st century farm crisis. In Holt-Jimenez E. and Williams J. (eds), *Land Justice: Reimagining Land, Food and the Commons in the United States*. Oakland, CA: Food First Books, 1–15.

Oliveira, G. (2013). Land regularization in Brazil and the global land grab. *Development and Change*, 44(2), 261–283.

Oliveira, G. and Hecht, S. (2016). Sacred groves, sacrifice zones and soy production: globalization, intensification and neo-nature in South America. *The Journal of Peasant Studies*, 43(2), 251–285.

Omestad, T. (2008, 25 June). American farmers try their luck in Brazil. *U.S. News and World Report*.

Paredes, M., Sherwood, S. and Arce, A. (2015). La contingencia del cambio social en la agricultura y la alimentación en América Latina. *Íconos – Revista de Ciencias Sociales*, 20(54), 11–25.

Romero, S. (2002, 1 December). U.S. farmers put down roots in Brazilian soil. *The New York Times*. Retrieved from www.nytimes.com/2002/12/01/world/us-farmers-put-down-roots-in-brazilian-soil.html

Stenner, P. (2008). A.N. Whitehead and subjectivity. *Subjectivity*, 22(1), 90–109.

Wesz Junior, V. J. (2014). *O mercado da soja e as relações de troca entre produtores rurais e empresas no sudeste de Mato Grosso (Brasil)*. Rio de Janeiro: UFRRJ.

Whitehead, A. N. (1978). *Process and Reality: An Essay in Cosmology*. New York: The Free Press.

5

AFFECTIVE STRUGGLES IN THE DESERT

Bringing water to bear on agriculture and food

Horacio Narváez-Mena

Introduction

The dry, cactus-dotted village of San Clemente, Ambuquí, located in the Chota Valley in northern Ecuador is an arid to semi-arid farming region, where rural families are dependent on irrigation for their food production and livelihoods. Once known for its cotton and coca leaf production during the pre-Colonial era, today large sugar cane plantations occupy the valley floor, with resource-poor families eking out a living on the surrounding barren, sand- and rock-laden mountainsides. For the people who live there, the continual search for water for their crops represents a daily crusade.

In this chapter, I look at a series of experiences in the village of San Clemente, where families provocatively transformed "a desert into an oasis" (Sherwood and Borja, 2009). While documentation of this activity addresses *what* happened – historically, in terms of the claimed source of ideas, funding, tubes, and sprinklers, it does not satisfactorily explore *how* such socio-material change came to be. The production of food in arid and semi-arid regions is not simply the sum of resources, but rather it is born from an endless array of human–nonhuman relations. My study is about how people generate the relationships necessary for access and utilization of water for food production. These entail endless, sometimes intentional and other times contingent human–nonhuman encounters (for example, among individuals and groups of people with watersheds, concrete, varieties of plastic tubing, and pumps) for managing climate, weather patterns and rainfall, ground and surface water, distribution devices, soil organic matter, plants, trees and other forms of land cover as well as planting dates, among myriad other possibilities.

People seamlessly mobilize knowledge, financial, material and biological resources to resolve the necessary relationships and their timing – what I refer to here as the "affective struggles" – for crop growth and food production. Inevitably, this requires involvement in a variety of social networks made up of family, friends and neighbors (and even enemies and strangers) as well as timely access to technical experts, authorities, and other institutional resources through a number of formal and informal means. With this activity, water, soil, plants, and sunlight, come together in a synchrony that gives birth to agriculture and food, enabling health, economy, and environment (as well as their antipodes). Here, I summarize the multitude of socio-biological-material relationships generated through people's affective struggle as "water for food."

Here, I go beyond the largely technical preoccupations that commonly limit irrigation studies to material outcomes to shed light on the underlying socio-biological-material relationships that ultimately catalyze the creativities and potentialities of water for food. In particular, I ask, how do affects underlie people's struggle for water in arid and semi-arid farming environments, such as San Clemente? And, what insights does an analytical perspective that takes affects into account provide to agriculture and food?

Methodology

Between 2011 and 2013, I carried out a total of six months of ethnographic fieldwork, involving six families located in the three neighborhoods of San Clemente, which previously had been identified as highly innovative in mobilizing water-for-food relationships (Sherwood and Borja, 2009). During that period, I lived in households, working in fields and participating in social activities. I joined farmers when irrigating fields and during other farm activities, such as manure collection, crop harvesting, and food preparation.

During fieldwork, the families studied generously allowed me to join them in their daily activities – from sunrise to sunset – and addressed my questions along the way. I focused my qualitative attention on their affective struggles.

Affective struggles

Upon my arrival, I noticed that the farmers I met regularly responded to my rhetorical "How are you?" with the phrase: "Here I am, struggling" (*aqui, luchando*). I came to hear this mention of struggle in a number of situations: informal conversations, interviews, meetings, and roadside conversations. At first, the term reminded me of the peasant and indigenous resistance struggles against oppressive regimes that is widely documented. Peasant struggles for water access, for example, have been widely documented, especially in terms of rights, governance, politics and gender (Boelens *et al.*, 2010; Boelens *et al.*, 2015; Hoogesteger, 2013; Roa-García, 2014). People in San Clemente, however, used struggle in a less antagonist way, which led me to this study. As noted by Adoum (1998: 53), to say that somebody is struggling can be a colloquial way of responding in Ecuador, as a way of people distancing themselves of the daily problems they face.

Nevertheless, in San Clemente, the application of struggle caught my attention because of the extremely dry desert-like climate in which people live, suffer, and thrive. Here, struggle is commonly not about creating distance from problems, but rather the opposite: facing life's challenges head on, in this case with regard to bringing scarce water to bear for food production and livelihood.

This is understood as a relational encounter among bodies. Such struggles, I argue, shape and are shaped by affect – the force by which bodies act upon one another, for example, in resolving the severe conditions of water and food scarcity in a semi-arid environment. In this study, I differentiate feelings and emotions from affects. As pointed out by Shouse (2005), affect is the most latent concept of the three. It cannot be expressed in words, and it is always pre-personal prior to a pre-consciousness (Massumi, 1995). "It is the body's way of preparing itself for action in a given circumstance" (Shouse, 2005) or "the power to act" (Negri, 1999: 85; Clough, 2007: 192). Quoting Carolan (2015: 319), "Affect can be taken to refer to a force or an intensity."

Hence, the aim of this chapter is to understand how the materialities around water provisioning (e.g., pipes, sprinklers and canals, etc.) are relationally bounded to everyday affective struggles in mobilizing water for food production. As Carolan (2015) and Arce *et al.* (this volume) argue, current agrifood studies generally miss a visceral emphasis on how food is felt, practiced, and performed. These authors emphasize that the nature of food production is relational, and this can only be grasped by looking at the affectivity of such relations (Carolan, 2015: 317). I apply this relational and affective perspective to examine the creative potential of bodies to exist in a context of water scarcity and general environmental hardship. People's affective struggles emerge out of the spatial and temporal relations between rural families and the highly contingent environments that involve them and in which they are involved. These relations lead to an endlessly creative transformation of the biophysical space where bodies live, perform, organize and operate. In continuation, I introduce my study site before going to explore the real time constitution of material and biological relationships associated with water for food in San Clemente.

Ambuquí: a context of water scarcity

San Clemente is located in the Ambuquí micro-watershed in northern Ecuador, where people have experienced water-related conflicts since colonial times because of unequal access to and distribution of the water that flows in mountain streams and the principal river. The watershed is located in the Chota Valley, about 140 kilometers north of Quito. Altitude ranges from 1,910 to 2,066 meters above sea level, and the valley is dry and hot, with an average temperature of 24 degrees Celsius. The terrain is mixed: flat areas, where agriculture is developed, surrounded by steep mountains with scarce plant cover. Soils in the arable areas are mostly sandy and rocky.

The village of San Clemente consists of three neighborhoods: San Clemente, El Lavandero, and Las Mercedes, with a population of approximately 180 inhabitants. The three neighborhoods share similar geographic and biophysical characteristics with climatic extremes and low water availability throughout the year. It receives an annual precipitation of 490 millimeters (mm), but it has an evapotranspiration of 1,149 mm, which represents an annual hydric deficit of 653 mm; making imperative the use of irrigation for agricultural production. The rainfall regime is composed of two dry seasons known as "summer" throughout the year. These go from June to August and from December to February. People rely on this seasonality for planning cropping activities. Nevertheless, this seasonality is not clearly marked since variations have occurred in the last decade. There are heavy episodes of rain in short periods of time, as well as extended periods of drought. This has increased the uncertainty for cropping activities; people say that they do not know anymore when to plant or when the rains will come.

In Ecuador, irrigation water is administrated by the National Secretary of Water, the entity responsible for giving legal rights to irrigation water called concessions. Because of conditions of water scarcity, in 1978, the provincial government started to build an irrigation canal – a project that was never completed because funding ended and because of limited commitment. The main capture system and canals are not fully functional due to frequent landslides that have blocked or destroyed them. According to the original plan the canal was intended to capture 146 liters per second from the Trapichuco creek and irrigate an area of 120 hectares. This water concession is divided among four users: two *haciendas* (large plantation operations; Irumina and Iruminita), the farmers' cooperative, Alberto Enríquez, in the village San Clemente, and the larger parish of Ambuquí. The neighborhood of Las Mercedes, which was formed later, did not receive any water concession and as a result, people only have small gardens that they irrigate with potable water. The neighborhood of El Lavandero received two legal water concessions, which are small compared with the one in San Clemente. One is of 2.5 liters per second, and the other of one liter per second. The first concession benefits two families, while the other is distributed among three families. These small concessions come from springs located in the highlands around Ambuquí that have been used for a long time and were legalized by the land reform in the 1960s and the purchase of land.

Arriving in San Clemente

Traveling from south to north across the Chota Valley on the Pan-American highway, the landscape is dominated by sugar cane, surrounded by steep slopes barely covered with vegetation. The sugar cane is green most of the time, thanks to an irrigation canal that brings water from Ibarra, the nearby city, to the *Ingenio Azucarero del Norte* (Northern Sugarmill). Approaching the exit towards Ambuquí, the view starts to diversify in terms of crops and the presence of communities scattered along the road. A number of resorts precede an exit that leads to a 1.5 kilometers long paved road up to Ambuquí. This village has benefited from the rehabilitation of this road in 2011. According to people from San Clemente, tourists from the nearby resorts visit the village to buy food and other local delicacies such as a wine made of the fig fruit *ovo*. Following the road and crossing the whole town of Ambuquí, which is about one kilometer long, an unpaved and rough road begins that goes to San Clemente and the communities upstream.

Despite the diversity that can be found in a relatively short distance between Ibarra and San Clemente (in terms of vegetation, infrastructure, ethnic groups, and weather conditions, to name some), it also presents contrasts with regard to access to resources, for example access to irrigation water by the sugar mill compared with the local communities in the valley. Or the use of water by the resorts for irrigation in an area surrounded by water scarcity. In the following section, I present examples of people's affective struggles in search of water for food.

"Informal" irrigation water

One day, after a meeting of a micro-credit lending group (*caja comunitaria)* established in El Lavandero, I asked the participants, "So, now that the summer [dry season] has arrived, what do you do?" Two of them responded in unison, "Continue struggling, what else can we do?" After a brief pause, they said that the amount of water in the *quebrada seca* (literally translated as "dry stream") had substantially decreased since the rains stopped. Farmers, however, had to continue producing food for sale and for themselves. One of the participants mentioned that he had a piece of land near the quebrada, which carries water to San Clemente and Ambuquí and, in spite of the fact that nobody in El Lavandero had legal rights to access that water, people had no other option than to use it. They even have established shifts or *turnos* (in Spanish) to distribute access to this source of water among themselves. There is not an official organization that controls use, but in their own words, "Everyone knows when it is their turn according to the local usages and customs," which date back to the period of the haciendas. In the past, El Lavandero was part of the large hacienda "Peñaherrera." The hacienda had access to the irrigation water that continues to the other two haciendas (Irumina and Iruminita) for irrigating the sugar cane plantations. Nowadays, however, small landholders, whose grandparents bought land from the hacienda, produce crops such as tomato, beans, and citrus without legal access to irrigation water.

One of the families I worked with used the water from the quebrada in the fields located along the riverside. In different conversations, they argued that they had the right to use that water because the hacienda irrigated those fields before they bought them. For this family, it was legitimate to use the water to produce food, even though their access to it was not recognized by the modern state or its laws. Farmers from Ambuquí and San Clemente instead have legal access to water and a *Junta de Aguas* (Water User Association) operates in Ambuquí and is in charge of managing the water that passes through the quebrada seca. Its jurisdiction, however, is limited to Ambuquí and San Clemente. Although the users from Ambuquí and San Clemente know that people upstream "informally" use their water, there have not been major conflicts. One of the main legal users is the Hacienda Iruminita, that makes little use of it. Thus farmers from El Lavandero feel it is unfair that they are producing food and do not have legal rights to that water.

To see themselves obliged to informally use water from other users as the only way of accessing water for irrigation is something that farmers from El Lavandero refer to as struggling. Although many feel they have historic rights to the water, using it without legal rights is generally perceived as stealing water. They joke about it; everybody knows who steals from whom, and also, as they said, they try not to cause any harm. Following their local usage and customs, they have managed to organize themselves around this informal (and illegal) way of accessing the resource through unwritten rules that they have agreed upon through their daily practices. Every time I touched on the issue, they responded that they were aware that they are stealing water from the users downstream. One of the participants in the meeting talked about how unfair it was for farmers from El Lavandero who produced food: "It is unfair to see how the water that passes in front of our noses is underused by the hacienda downstream. They [the hacienda] have water every day, but they are not producing anything. They are misusing the water with some sugar cane they have in there."

Alfonso, one of the farmers, referring to a bridge that the provincial government had offered to build, added, "Once we get this bridge, I will continue the 'struggle' for the water, I will investigate what happens with the water that goes to the hacienda. I will find out whether they have a concession or not and see if we can get at least two liters per second out of all the water that goes to the hacienda."

"Stolen water" is used for both own consumption and the market. People get away with it and continue producing. Nevertheless, it is difficult to live and work so far from the marketplace in Ibarra and be informed on commodity prices. Overcoming this limitation requires direct access to people in the know, such as local vendors and truck drivers who regularly travel to commercial centers. One day while leaving El Lavandero, I met a farmer on the road who was also traveling to Ibarra. He described how he managed to keep informed on how he got away with stealing water as well as how he kept up to date on market prices:

> I have beans, sweet potato, cassava, that is the advantage of living in the countryside, you have food ready [tener a mano] and you do not have to go out to buy it. Right now, I am going to Ibarra to see whether I can sell tender, young beans (*frejol tierno*) that I have. It seems, it got a good price! I am going to the wholesale fresh market. I have friends there who put me in contact with traders and middlemen and they come directly to the field to pick up the beans … All the land that you can see in front of the communal house is mine. I don't have irrigation water and I have to steal. You have to be smart to steal water. I only steal the minimum necessary, in such a way that people downstream don't even realize you are stealing their water. There are people that steal a lot though. Then, people downstream realize it and come and cut off the irrigation water.

Another farmer described the difficulty of producing without access to irrigation water:

> In the dry season it is very difficult to produce here, farmers do not have anything in this season. The beans, the peas, are crops that easily get diseases, so farmers do not cultivate them. If you start cultivating beans now, they will not come out, if they do they will be ugly, you will lose your harvest. It is very expensive to produce in summer (dry season). There are people who take the risk though, but they have [irrigation] water. They can get a good price selling the frejol tierno.

There are cases where the informal use of water is the only source for irrigation. One farmer said, "To steal water is not something to be ashamed of, neither it is a sin, people survive from it and I am one of those." Another farmer told me the story of how she and her husband had always used the water of the canal even investing money in irrigation systems despite not having a legal concession:

> We bought the pump with credit from the cooperative Cochapamba [a credit cooperative from a nearby community]. We have two reservoirs. We take the water that comes from the irrigation canal with the pump to bring it to this reservoir [one next to the house]. Then we carry up the pump and pump the water again from here to the other reservoir on the top of the field. From there, we irrigate by drip irrigation all the avocados and the other trees. That way, the plants have become greener.

Alfonso and Olga's transformation

Alfonso was one of the first people to arrive in El Lavandero. When he was still a child, he arrived accompanied by his grandfather who at that time worked at the Hacienda Irumina. They came from a community called Peñaherrera located higher in the mountains upstream from Ambuquí. As time passed, Alfonso's grandfather bought a large plot of land (about 15 ha) from the hacienda owner with the intention of leaving it to his children who had also decided to move to El Lavandero. When Alfonso and Olga married, they received a piece of land from Alfonso's grandfather and started to live in El Lavandero. Since water was always scarce in the area, during an interview, I asked Alfonso to tell me how he got legal irrigation water from a spring on his farm, and he responded:

> My grandfather struggled to get water since we arrived in El Lavandero. The water that I now share together with my two uncles was from my deceased grandfather. When he died, there were other sons that wanted to get the [legal] concession [of the water from the spring]. My two uncles and I had to fight to defend the water because we invested time and money to get it. When we were going to pipe the water and bring it to the farms, we asked the others (the other uncles) to join us but they didn't show interest and didn't help. When my grandfather died, we had to enter a law suit to keep the water because the others also were claiming their right to that water.

> Nobody can make us be without water, in particular since we have been using it and we have fought to keep it and bring it to our farms.

The concession that Alfonso's grandfather obtained was located about five kilometers from El Lavandero and was channeled in an open canal to the fields. Piping was uncommon. Water was not available all the time, and when it was, they had to take full advantage of it, even to irrigate during the night. Furthermore, the water was divided among three family members, Alfonso and two uncles. Access to irrigation water generates interest from different parties in this area. From the above description, it is clear that accessing irrigation water was a struggle, not only against the environmental conditions of drought, scarcity and natural disasters but also between people, in this case, among members of the same family. Regarding this issue, Alfonso mentioned that it was particularly difficult to deal with a conflict like this with his family. In fact, it has caused some divisions between them.

Despite such conflicts, obtaining water was a key step in transforming the farm. They started to produce some annual crops typical of the area (beans, bell peppers, and tomatoes). Nevertheless, it was not enough to make a living. Alfonso mentioned that he was planning to migrate to Spain to get more income to support his family. At that time, maintenance was being carried out on roads in El Lavandero using machinery. The operator suggested that they construct reservoirs for water and Alfonso took the opportunity to build one on his farm.

> The tractor operator used to work [on the roads] until 1 pm. He wanted to work extra hours to make a little bit of extra money. He offered to build the reservoirs for us, that is why we have the reservoirs; he helped four people.

Having the reservoir, Alfonso and two other uncles decided to pipe their water concession from the source, to reduce loss from highly sandy soils and evaporation. Alfonso was not satisfied. He once said in front of his children that he was ashamed of being a farmer because he could not provide them with enough commodities to live. His struggles were directed towards getting food for his family and getting water was one step forward. In a way he was being affected by the environmental and economic conditions, and in turn, he was affecting his surroundings by looking for alternatives to transform the farm. Another organization invited Alfonso to participate in a field trip to visit a farm with water harvesting practices. Alfonso had shown particular interest and was willing to try out new things in his farm after the trip, which called the attention of the technicians from the NGOs who work in the area. He remembers:

> We visited a farm that cultivated around 10 hectares only harvesting rainwater. That was an impact for me! I told myself, "I even have a small spring, and they can do it only with rain water …" Then, the "ingeniero" asked me: "What do you think?", I told him: "Very good! I will follow this,

> even little by little." A week passed and the "ingeniero" came to tell me that they wanted to implement a pilot project here, the first one.

The field trip was an eye opener for Alfonso and Olga about how to farm:

> I was totally convinced of changing from the other way of doing agriculture that we had here. First, regarding water, we did not have the knowledge on how to optimize water. Maybe, with what we had, we could produce only a little bit. Secondly, the traditional crops were not profitable anymore. In those times, if the summer arrived, then you lost and had to migrate to the city or other places to look for other sources of income. But that was the past; today is different.

Alfonso and Olga started to produce alfalfa under sprinkler irrigation for raising guinea pigs. He also got fruit trees including citrus and avocados and watered them using drip irrigation. His farm became one of the agro-ecological farms promoted by the local NGO AGRECO. In El Lavandero, Alfonso and Olga were the first to adopt the new technologies on their farm. According to Alfonso, people in the community were skeptical about installing drip irrigation, even his uncles told them that he was wasting his time. This space, however, opened up other opportunities for Alfonso, both within and outside the farm. After looking at how the farm was becoming greener and greener, their neighbors also started installing irrigation systems and asking Alfonso for assistance.

The field trip created new relations between the family and the organizations but also between the family and their farm. Implementing new practices meant acquiring new knowledge to put them into practice. The relationship with water had changed from being a scarce resource to be stored to something to be managed. On a number of occasions, Alfonso and Olga mentioned that in the past they did not have enough water for planting, but after transforming their relationship with water, they now do not have enough land for production. Nevertheless, this is not the end and is far from being perfect. The transformation of the farm is a continuous and never-ending process. In the everyday struggles they achieve a transformation.

There are other instances where struggling means dealing with daily farm activities. During a field visit, I found Alfonso fumigating a field of beans. He was spraying insecticide with a backpack pump and wearing a big plastic bag on his back to protect himself from the insecticide. He was not wearing a mask, glasses, gloves, or any other sort of protection. I waited until he finished fumigating and then I went up to greet him:

> "Good Morning Don Alfonso, how are you today?" He extended to me his forearm instead of his hand, because he was fumigating, and replied: "Here I am, struggling, the *ingeniero*. You know, there is this *palomilla*[1] that is attacking the beans," showing me the back of an infested leaf. He continued: "It is the

first time I have fumigated these beans. The palomilla came from another field where I also had beans, but I have already harvested it. Olguita is threshing it. There are not many other fields with beans around, so the palomilla doesn't have anywhere to go and comes here." I then asked him why he was planting this crop now that the dry season started, and nobody else is doing it: "I saw that my uncle had the same variety and got a good production, so I went to Ibarra to look for the seed. This variety is drought resistant; if this was another [variety of] bean, it would be yellowing now and wouldn't resist the drought. This one doesn't need much irrigation. Despite the 'summer', these beans are going well. Don Galo [his neighbor] also has this one [bean crop]. If it were raining, these beans would be very good, even better."

Discussion

What is the struggle for water for food? Referring to the affective turn in agrifood studies (Carolan, 2015), this research explored the endlessly creative processes involved in living and being, in particular with regard to acquiring water for food. This inquiry brought to light at least three interactive processes: (1) the hidden (or neglected) human–nonhuman relationality of water-for-food, (2) the affective dimension of the struggle and (3) the transformative power of such relationships.

A relational perspective puts forward the notion that relations between entities are more fundamental than the entities themselves (Wildman, 2010; Blanco *et al.*, 2016). As I have mentioned earlier, food is the product of relations rather than the sum of resources. In the cases presented, both social and socio-material relationships influence the way people get access to the resources they need, in this case accessing water to produce food. At the community level, creating relations within the community has been essential for establishing informal rules for water use. Creating relations outside the community, on the other hand, has also been important for obtaining resources, for example, for bridge construction. This becomes relevant when considering food production. Its actuality can only be grasped by looking at it as a process composed of relationships. The construction of the bridge might have nothing to do with accessing water, but it has to do with facilitating the transport of irrigated products to markets. Also, it is an opportunity for the community to become more visible to the authorities. Such relationships also are reflected at the family level. As in Alfonso and Olga's case, family relations have been central in obtaining water concessions. In particular, Alfonso's ability to mobilize his networks and gain access to technicians in NGOs and subsequent leadership in a project and in his neighborhood enabled him to obtain specific resources for his farm, family, and friends, thereby strengthening his socio-material potential.

To argue for an affective dimension of water for food production is to take into account how it feels to produce food (Carolan, 2015). By looking at the practice of struggle, it becomes evident that affects are not merely abstract but tied to and put into motion through relations among people, objects, biology, plants, and so on.

They are connected but forgotten in the oversimplification of development practice. Although I have previously made a distinction between feelings, emotions, and affects, in practice, they are difficult to separate. From the cases presented here, it seems that people use struggle as a way of expressing what they feel (feelings and emotions) with regard to living in the environmental or economic conditions (e.g., sadness, anger disparity, unfairness, and shame) as a way of preparing themselves, to respond to such circumstances (affect). Affect becomes the trigger that reconfigures, redirects, or influences actions they take to overcome such conditions (the power to act), for example how Alfonso dealt with getting water concessions, leading the construction of the bridge, or fumigating against a pest in his crops.

When looking at such specific and unique cases, a question that arises is how these examples may be useful in promoting innovation for solving problems where water is scarce. From these cases, changes certainly have occurred (i.e., the diversity of crops and growing "greenness" of Alfonso and Olga's farm is visibly different from their neighbors). They might be located, though, in the margins of the actual potential and possibilities of low, external input, intensive, smallholder family farming. Carolan (2015) calls these localities "incubation spaces" and argues that such experiences need to be addressed through policies to ensure that they are replicated. A field trip, a loan group, or even a drastic climatic event can trigger such sites. In this study, looking at affective struggles has helped to identify what can potentially be incubation spaces in this community. One such place could be a farm where the conditions of scarcity and a history of struggles over resources have triggered its transformation.

Conclusions

This chapter opened with a reference to Carolan's (2015) affective turn and how present-day institutions of development practice, and in particular water management for food production, in their obsession with getting people to change through the promotion of pre-imagined, universal technological packages, such as water harvesting and micro-irrigation systems or community credit schemes, are effectively blinded to the realities of people's social bodies. Through in-depth study of family and community-level practices, development initiatives interested in rural transformation through water would stand to benefit by paying special attention to people's affective struggles – understood as not merely an abstract battle but also their highly inspired and self-organized battles in favor of certain individual and collective tastes, cares, textures, and practices associated with alternatives to the status quo in water management for food production.

Affects are entangled in the relations among bodies of all sorts – thoughts and imaginations and even human–nonhuman interactions. Affective struggles are visible in rural relations and their associations brought forth in the need and inspiration for water and food, reconfiguring, redirecting, and influencing potentialities and capacities to act. This is reflected in family practice, transforming people and their organization as well as the color, shape, humidity and even the emotions of the physical landscape. In the process, for example on Alfonso and Olga's family farm, what was once a desert, became an oasis! The experience here shows that alternative foodscapes are not just possible (Giller *et al.*, 2009; Andersson and Giller, 2012), but they actually exist. At present, this richly creative, socio-technical-biological dynamic largely lies outside the institutional imagination of development practice and, as such, it is either neglected, lost or strategically displaced as a possibility for public debate and a resource for social change.

An object–subject interaction provides the opportunity for looking at water–soil–crops–food as a relational process. Given the substantial obstacles to working with development institutions of science, the state and other agencies, a question is raised over how development initiatives can work in such a way that the care, nurturing, and "greening" of landscape for more healthy, sustainable food production become more tasteful.

Certainly, part of the answer rests in the undeniable diversity experience in San Clemente. Nevertheless, as emphasized elsewhere (Mueller, 2006; Cordoba and Jansen, 2015), rural development is highly political and politicized, especially by technocratic networks that champion agricultural and food modernization. As argued in Sherwood, van Bommel and Paredes (2016), an "affective turn" in development practice would require special attention to the very issues and concerns that have displaced affect in the first place. Rather than seek a "fix" in the present-day techno-centric institutions of development practice, a "bypass" may be necessary for the required attention to the multitude of relations and associations of the daily affective struggle involved in bringing water to bear on agriculture for food.

Acknowledgements

The research reported here was part of a Dutch-financed NWO–WOTRO Programme on "Catalyzing Positive Deviance in Agriculture and Food."

Note

1 Also called *mosca blanca* or whitefly (*Trialeurodes vaporariorum*).

References

Adoum, J.E. (1998). *Serie magister: Ecuador: señas particulares (ensayo)*. Quito: Eskeletra Editorial.

Andersson, J.A. and Giller, K.E. (2012). On heretics and God's blanket salesmen, contested claims for conservation agriculture and the politics of its promotion in African smallholder farming. In J. Sumberg and J. Thompson (eds), *Contested Agronomy, Agricultural Research in a Changing World*. Abingdon: Routledge, 22–46.

Blanco, G., Arce, A., and Fisher, E. (2016). Intersubjetividad y domesticación en el devenir de una región global: la territorialización del salmón en la Patagonia chilena. *Íconos: Revista de Ciencias Sociales*, 54, 125–144.

Boelens, R., Bustamante, R., and Perreault, T. (2010). Networking strategies and struggles for water control: from water wars to mobilizations for day-to-day water rights defence. In R. Boelens, D.H. Getches, and J.A. Guevara-Gil (eds), *Out of the Mainstream: Water Rights, Politics and Identity*. London: Earthscan, 281–306.

Boelens, R., Hoogesteger, J., and Baud, M. (2015). Water reform governmentality in Ecuador: neoliberalism, centralization, and the restraining of polycentric authority and community rulemaking. *Geoforum*, 64, 281–291.

Carolan, M. (2015). Affective sustainable landscapes and care ecologies: getting a real feel for alternative food communities. *Sustainability Science*, 10(2), 317–329.

Clough, P. (2007). Introduction. In P. Clough (ed.), *The Affective Turn: Theorizing the Social*. Durham, NC: Duke University Press, 1–33.

Cordoba, D. and Jansen, K. (2015). Realigning the political and the technical: NGOs and the politicization of agrarian development in Bolivia. *European Journal of Development Research*, 28(3), 447–464.

Giller, K.E., Witter, E., Corbeels, M., and Tittonell, P. (2009). Conservation agriculture and smallholder farming in Africa: the heretics' view. *Field Crops Research*, 114(1), 23–34.

Hoogesteger, J. (2013). Movements against the current: scale and social capital in peasants' struggles for water in the Ecuadorean highlands. PhD dissertation, Water Resources Management, Wageningen University, the Netherlands.

Leitch, R. and Day, C. (2000). Action research and reflective practice: towards a holistic view. *Educational Action Research*, 8(1), 179–193.

Massumi, B. (1995). The autonomy of affect. *Cultural Critique*, 31, 83–109.

Mueller, S.D. (2006). *Rural Development, Environmental Sustainability, and Poverty Alleviation: A Critique of Current Paradigms*. Working paper no. 11. UN Department of Economic and Social Affairs.

Negri, A. (1999). Value and affect. *Boundary 2*, 26(2), 77–88.

Roa-García, M.C. (2014). Equity, efficiency and sustainability in water allocation in the Andes: trade-offs in a full world. *Water Alternatives*, 7(2), 298–319.

Sherwood, S. and Borja, R. (2009). Creating an oasis in the desert. In C. de León, B. Douthwaite, and S. Álvarez (eds), *Most Significant Change Stories: Global Challenge Programme on Water and Food*, CPWF working paper no. 3, 82–85, the CGIAR Challenge Program on Water and Food, Colombo, Sri Lanka.

Sherwood, S., van Bommel, S., and Paredes, M. (2016). Self-organization and the bypass: re-imagining institutions for more sustainable development in agriculture and food. *Agriculture*, 6(4), 66–84.

Shouse, E. (2005). Feeling, emotion, affect, *M/C Journal*, 8(6). Retrieved 4 April 2016 from http://journal.media-culture.org.au/0512/03-shouse.php

Wildman, W.J. (2010). An introduction to relational ontology. In J. Polkinghorne (ed.), *The Trinity and An Entangled World: Relationality in Physical Science and Theology*, Michigan, IL/Cambridge: William B. Eerdmans, 1–14, 55–73.

6

PEOPLE, COWS AND MILKING MACHINES

Public policy and intersubjectivity in Ecuador

Nicolas Vasconcellos

Introduction

This chapter draws on the experience of peasant farmers to describe and analyze how the introduction of milk production technology transforms relationships and creates new subjectivities. This technology is seen as part of the Ecuadorean government's plans for improving agriculture through technology-based interventions. The Ministry of Agriculture, Livestock, Aquaculture and Fishing (MAGAP in Spanish) targets peasant farmers as beneficiaries of technical knowledge, creating space for a knowledge encounter between competing regimes of practice.

During 2015, *ingenieros* (agricultural technicians) from MAGAP arrived in rural villages across the country with milking machines, artificial insemination, and other technology that was intended to improve the quantity and the quality of milk production. Small-scale producers started to diversely utilize these machines, initiating a process of localized adaptation. To get a better grasp on the human–milking machine interface and to shed light on agricultural policy in practice, I conducted an ethnographic study.

Following Stenner (2008), the farmer–ingeniero encounter over agricultural transformations brings to the fore a division between subject and object as cognitive and relative terms that have guided and contributed to pre-designed technological intervention. This division that presumes the separation of the world in people and things is an important cognitive frame with which to regard change and give meaning to the "modern world." In this frame, policy formulation and intervention are almost always the institutional property intended to work on people and in things, be it the body of an "underdeveloped" actor that suddenly enables progress, or increased productivity to address a situation of poverty. Such activity is an irrefutable manifestation of the logic of rational efficiency and institutional responsibility in contemporary society in places such as Ecuador.

Institutional policy interventions in favor of development and progress speak the "truth" of modernization. The purest division of object and subject had been uttered by nation-states' bureaucrats and institutional experts and by "some of the modern things" themselves, which usually are presented without the distorting filter of human experience or the reflective instance that change also generates. The occasion of experiencing change and transformation question the above-mentioned division between objects and subjects to understand and transform the world. Competing regimes of practice are the result of this epistemological polarity, which explains why state representatives and peasant farmers frame interventions in rural studies as a situation of reversal and replacement. Thus, a presentation of a technological intervention, where animals, machines, and peasant farmers are depicting intersubjective relationships of affect is a case threatening to turn the tables on the predominance of expert knowledge in rural development. Nevertheless, to acknowledge the nonhuman as a potentially active agent in social change involves an ontology that questions the separation of the object and subject. In practical terms, it demands awareness of the possibility of how worldly things (both animal and machine) may come to constitute the peasant "reality" during moments of technological intervention and thereby occupy the rural. This ontology takes the significance of people's experience in how the world is made seriously.

In this case, I explore intersubjectivity tied to the introduction of a milking machine, cows, and peasant farmers as one of the fundamental features of rural experience and institutional intervention (Whitehead, 1933/1935). Following this orientation, I suggest that it is in the distribution and involvement of affects where we can identify the social ground that is conducive to change in existing regimes of practice. This case may unsettle views about the "nature" of agency and the importance of people, but also machines and animals.

Here, intersubjectivity refers to the emergence of affect when things are not understood as subject and object, but as entities which gather together in a new corporality. This corporality embodies the farmer, the animal, and the machine in a process where the agency is distributed equally in these three entities of the assemblage. The glue of this happening is not just the cognitive knowledge of the farmer but also the embodiments of her or his experience, including the affect for the cow and the understanding of the potentials and limitations of the milking machine.

Nevertheless, a separation between matter and policy can dilute how actors may grapple with subject–object interfaces in practice. When treated as mere object, an entity that is the object of a particular policy tends to appear as if it were passive and without affect. Although, the materiality of the rural space, not unlike animals and machines, is independent from any cognitive policy formulation, some specific material properties, such as the noise of the milking machines or the nervous disposition of a cow, must be taken into account in order to recognize "rural" intersubjectivity. The latter is defined as the property of people and things to generate new world corporalities, which constrain policy intervention to the properties of rural materiality. This materiality stakes its claim in the gatherings by which the rural world is made and in the alliances people and things create or depict therein, such as the material zone of transition forged during a policy intervention.

As emphasized in Arce *et al.* (2015), peasant-farmer exposure to modern political and social situations, such as transnational economies, cross-cultural interactions, and mass migration, have contributed to a more diverse and mixed view of what is contemporary reality. Here, the notion of experience detaches itself from classical, collective identities of communities and the nation-state. Thus, political institutions, such as those found in the nation-state, are questioned more often by people, as they are not the only valid political configuration in today's highly fluid world of things, ideas, and people. As argued by Braeckman (2008), in this context political life is understood as increasingly lying beyond the formalized bureaucratic confines of the nation-state, instead resting in the less formalized, sub-political spaces of people's everyday practice in civil society.

Here, I work with milk production in the context of Ecuador's pioneering, if ambitious policy of "food sovereignty" to observe the introduction of technology into rural associations and resulting knowledge encounters among ingenieros as government agents, peasant farmers, and cows. In particular, I am interested in exploring the intersubjectivity that results from the gatherings when actors dismantle, (re)construct, and constitute materialities, thereby crossing through boundaries that constitute a new particularity (Deleuze and Guattari, 1988), in this case in reference to the corporality that emerges from the interrelationships among people (technicians and farmers), cows, and milking machines.

The arrival of food sovereignty in Ecuador

Representatives of the Ecuadorean people wrote food sovereignty into the country's 2008 National Constitution as a strategic means of increasing food production for its neediest populations, while ensuring that food remained in the hands of the public. The concept is understood as the right of people to healthy and culturally appropriate food produced through ecologically sound and sustainable methods as well as the right of people to define their own food and agriculture systems. Ecuador's 2010 Law for Food Sovereignty (LORSA) sustains the "promotion of sufficient production and adequate conservation, exchange, transformation, commercialization and consumption of healthy, nutritious food preferably grown by small, micro and medium-sized peasant producers."

For this, the state publicly commits to investments in scientific and technological research in agro-food processes. In addition, it ensures investments in "extension" programs, aimed at "transferring" technology between centers of knowledge and farming practices through means of dialogue and exchange between ingenieros and small and medium-sized production units. Finally, the state commits to safeguarding the ability of farmers in rural communities and indigenous nationalities to preserve and promote sustainable water, soil, and biodiversity management, thereby guaranteeing the necessary conditions for maintaining, protecting, and developing collective traditional know-how, scientific knowledge, technology, and genetic resources for assuring stable food production.[1]

Nevertheless, smallholder peasant families struggle to use government programs in their favor in order to connect to a market that pays a fair price for products, thus helping to improve not only agricultural production but family well-being as well. On top of this, the government established tighter policy reforms regarding short-term labor contracts, leading to an increase in daily wages, mandatory pay rises, health care and retirement provisions. This important welfare orientation has a substantial increase in labor costs, which practically doubled from one day to the next. While such policies have benefited marginalized laborers greatly, the law ignored that people in rural areas of Santo Domingo only have part time contracts, so that they can go back home and work on their own crops. In addition, the law demands increased pay during weekends and holidays, making the hiring of labor impractical for many farm owners.

The government has sought to establish new political spaces for the sake of greater social participation and civil control over society. The Plurinational and Intercultural Conference of Food Sovereignty (COPISA), which works for the National Assembly and is composed of eight representatives from civil society, was created to generate civil society participation and debate processes on the formulation and implementation of policy, plans, programs, and projects related to food sovereignty (COPISA, 2014). Nevertheless, the policy proposals generated until now are kept on hold by the government, which brings to fore that the new "spaces" may be autonomous from one group of actors, while being partly or totally controlled by others (Brock, McGee, and Gaventa, 2004). This situation places food sovereignty at risk of being manipulated by actors in the state with a mask of civil society participation (Arce, Sherwood, and Paredes, 2015), while perpetuating the long-standing modality of institutional nationalism as a style to achieve political development.

Methodology

The experiences presented pertain to the Agricultural Association of Cristóbal Colón (AACC), a precinct[2] in the province of Santo Domingo in the northwest of Ecuador. This region is characterized by monoculture export crops such as pineapple, cacao, and palm oil as well as cattle and poultry production for national markets. While many financially important products come out of this province, MAGAP is largely absent and disconnected from the everyday realities of the peasant-farmer livelihood. This situation generates knowledge gaps, which are important to explore in order to understand the materialities that are created by the policy space and the embodiment of these technological interventions, which emphasize the local entanglements between "rural matters" and the policy meaning of the intervention.

Due to a major drought and ensuing economic crisis in southern Ecuador in the 1960s, much of the present-day population of Cristóbal Colón had migrated from Loja to Tsachila lands. These people became connected with the advent of urban-based markets, goods and public services during the 1980s, and it is from this period onward that farmers began to visit Santo Domingo and Quito more frequently. The arrival of roads allowed a growing number of rural families to educate their children outside the precinct, and this reinforced the aspiration of children to seek opportunities outside agriculture. In turn, this opening up of the rural area brought influences from urban migrants, who in keeping ties with their family and even returning to their rural villages drew on their relationships and experiences in the city to transform the *campo*.

Through ethnography, I explore how the peasant farmer exists in a multiplicity of production practices, where subsistence and food consumption generate new knowledge. Within this constant struggle to distribute state policies of modernization and peasant farmers' constant actualizations lies Ecuador's ambitious and controversial food sovereignty policy. The state leans towards the achievement of technocratic programs, thus remaining blind to rich peasant farming experiences and how the present and future ultimately depend on rural experience and pragmatic relationships, as brought forth through the creativity of intersubjective assemblages.

I document this experience through fieldwork that took place over one year during 2015 and 2016, involving semi-structured interviews, participant observation, and life histories. Individual participant observation was carried out with 15 producers, including 40 semi-structured interviews. The research is mainly situated in the households and communities of the AACC members, who live about 40 minutes from the town of Santo Domingo. I also participated in a number of MAGAP sponsored events, which took place in surrounding villages and usually included farmers from neighboring provinces.

The peasant farmer as central agent in food sovereignty

During the year that I was in Santo Domingo de Los Colorados, I lived with a peasant farmer and his wife (Don Wacho and Mrs. Dona). Don Wacho was born in Cariamanga, Loja (650 km from Santo Domingo). His father worked for a large-scale cattle operation, where his family resided. When Don Wacho was six years old he moved to the city of Machala (200 km from Loja). In 1967, he married Dona, and the couple remained in Machala for 23 years, where Wacho worked as a refrigeration technician.

During that period, Don Wacho held a number of leadership positions in his community. He was president of his neighborhood, and he led a number of initiatives to organize people and make demands for basic services from the municipality. During that time, he always saw great strength in community organizations, and he believed in the power of association as a means of achieving fairness and justice for all the members of a given group. However, when his father-in-law fell ill, Don Wacho moved to Cristóbal Colón, Santo Domingo to help take care of his farm. After several years, his wife came to join him on the farm, and following the death of his father-in-law, the couple inherited the farm and found themselves working the land for a living.

At present Don Wacho is the manager of the Agricultural Association of Cristóbal Colón, to which he belongs since arriving in the precinct in the late 1990s. As he once said, "For me this association is a hobby. As I told you before, I come from a community area where money does not make me who I am … where I have to look for people so that we can all benefit from and not harmed by government policies" (Fieldnote 27 April 2015). Since childhood, Don Wacho has been moving between urban and rural areas and vice versa. These experiences led to a "mutation" of the peasant that was exposed to constant changes that often comprise rapid and unpredictable re-assemblages of the recursive properties of entities – a situation that redraws existing boundaries and promotes the emergence of new social forms (Arce and Long, 2000).

Don Wacho and the 43 members of the AACC have acquired not only their own property, but through their advocacy they managed to capture the attention of the ingenieros at MAGAP, which offered them support through state-sponsored training, marketing of their products, financing, and equipment for the development of infrastructure, such as the creation of a milk collection center. As individuals, peasant farmers have access to certain state support programs, but when this fails to satisfy their needs or interests, they mobilize their farmer association to strengthen their voice and make claims and demands. This transition from individual to association means that the borders of the state can be effectively challenged and renegotiated. In this way peasant farmers manage critical interfaces between a variety of local situations and social and political realities in which they live and operate.

This activity can become disrupted when MAGAP targets an association with a technological innovation program and a team of ingenieros arrives in the community, causing peasant farmers to mobilize in the hope of managing the situation in favor of their interests. The farmer or the association may not show its concern or disinterest in the particular proposals, for example in the promotion of a certain suite of technologies. Instead, they will seek explanations and explore pros and cons. A manifestation of this can be observed when the MAGAP brings technological programs to agricultural associations and the peasant farmer hesitates to put this technology in use, however, without explicitly refusing or accepting the technology in question. This process takes place in interventions stimulating local institutions for development, such as a community bank, the repairing of infrastructure, a vaccination campaign, or the conformation of an organization.

In this case, the peasant farmer is directly involved in negotiations at the same time that the state is implementing its policy. Hence, in terms of the actualization of a local livelihood, at the moment of implementation there is no clear boundary that separates the competing agents from operating on behalf of state institutions (Beck, 2006). It is in these situations where the peasant farmer must find a way to redefine her interests and work to re-establish social and political interrelations.

The main cause of this trespassing of state boundaries is the social dynamics of civil society and its individualization processes[3] (Beck, 1993; Braeckman, 2008). I go on to present an example of a conflict between MAGAP's ingenieros and peasant farmers.

A knowledge encounter

In the case of the producers, whether using the milking machines or following MAGAP insemination protocols,[4] acceptance of change depends upon its resonance within established ways of doing as well as the local actors' ability to manage competing influences and negotiate (Arce, 1993). In a visit I made with Carlos, a livestock technician from MAGAP, to see if producers were properly using MAGAP's milking machines, I asked him what should be done when a cow became aggressive while the machine was working. His response was that the farmer should change the cow for a different one. This was a curious point of disagreement with producers of the association, since the technician insisted on a certain cross breeding, but the producers disagreed with the breeds offered by the technicians. The technicians tried to improve the breeding of local animals by teaching producers to use artificial insemination with MAGAP-selected breeds. One justification for artificial insemination was avoidance of the need to keep a bull, as bulls commonly cost US$ 1,500 or more! Nevertheless, when I visited farms, essentially everyone had one bull, if not two. When I asked why, they responded that one has to constantly see if the cows are on heat, in order to inseminate them at the right moment. In their experience, insemination, artificial or otherwise did not work, if the cow coming on heat was missed. This clarified to them, it was better to have a bull, and let it impregnate the cow at the right moment, as it was certain that the cow would remain pregnant and in continual production.

In a following visit to the AACC, Carlos said that MAGAP would not be receiving more heat-inducing hormonal implants for cows, so they had to work with a cow's natural heat. This involved training someone in the community who could potentially inseminate producers' association livestock. When I asked about the reasons for not using the hormonal implants, he explained:

> People do not want to change. They are loath to change and that has to be fought with. Besides, we have been working with other producers using the syncing procedure [implant heat inducers to make herd heat uniform], and the procedure was not so effective, because the rate of pregnancy with the implant is very low. A best case scenario is around 55 percent, so people do not want to hear anything about it. Maybe it was not very well explained to them, and their expectations were higher. Also there are producers who say that after the cows were synced and inseminated by MAGAP, they did not get pregnant and have not gone into heat again. All sorts of things have been seen and heard, and those things may be possible. Besides, when this is done

> as a private service [with business farmers], you prepare the animals – get them to be well fed and in good body condition. But here [with the agricultural association], it is not always possible to do it with all of those details. And the success of syncing rests on those details.
>
> *(Fieldnote 23 April 2015)*

Such MAGAP interventions in the lives of peasant farming give space to the creation of a body of knowledge, which is presented as part of a homogeneous body of "best practice." In Cristóbal Colón, this clash of agenda and worldview inevitably generates conflict between ingenieros and local farmers. As food sovereignty policies are directed almost exclusively at peasant farmers, field technicians of the ministry, as agents of the state, operate at the cost of local knowledge.[5] In this situation, the concerns of the technician largely reside in the efficiency of their intervention, but also with the task of popularizing science and technology as well as modernization (Arce, 1993). The common determination of such assessment is the "lazy" or "laggard" farmer.

Milking machines: the emergence of a new intersubjectivity

During my first visits I was accompanied by María, a field technician from MAGAP. By the time I started my fieldwork, the AACC had received ten milking machines. According to María, the machines had lain in storage for two years before being given to the farmers, and as a result some machines had mechanical problems due to a lack of use. A technician from the Technological Innovation Department told me that these machines had been part of a credit program from the National Development Bank (BNF). This program owned machines that never reached farmers. María said each machine cost US$ 4,500 each, but since the machines were part of a program which had already bought them, producers who wanted a machine under the new project only had to pay US$ 600 each. The funds were used to repair the machines and make them functional again. Therefore, Don Wacho spoke to the 16 milk producers of the association about the availability of the milking machines. Ten were interested but only six could pay immediately, so Don Wacho proposed that the remaining four take small portions of their milk payment to cover the cost of obtaining their own machines. This suggestion was accepted.

The pre-established idea of field technicians that peasant farmers are subjects who would automatically adopt any program brought by MAGAP is a view that bypasses the fact that peasant farmers must assimilate any technological innovation according to their materiality and experiences. The case of the milking machines, and their aim to improve milk production, brings along new bovine spatio-temporalities.[6] This is a new intersubjectivity which is co-produced by the encounter between policy, technologies, and the material characteristics of the existing milking practice. This encounter enables animals to express their already existent, yet taken-for-granted intersubjectivity (Bear and Holloway, 2015) within the process of innovation.

During the 12 months I visited the producers of the association, I noticed that even when they had already received the milking machines, only two or three used them. The rest did not, explaining that they first needed to set up a specific place to start using them; others said that the loud sound of the machines scared their cows and made them aggressive. During that year the MAGAP ingenieros visited to make sure the machines were being used. Otherwise, they would confiscate the machines and give them to other producers. (I was told that the intended beneficiaries were commonly large-scale producers who had friends in the DPA, even though in some cases they already had machines.)

While not many of the farmers made use of the milking machines, those who did argued that it was a daily struggle to get their cows to adapt to the regular milking routine with the machines, because of the loud noise. One morning during one of my early visits to one of the producers, I started walking towards the cattle pen and I could hear the loud sound of the machine as well as endless bellowing from the cows. Don José was tying up his cows to the chute[7] in preparation for milking, when he said:

> The cows are stubborn today. They don't want to give milk. They are retaining the milk when we use the machine; we have to repeat the milking session because they are retaining it. We had to milk one twice with the calf[8] to get all of the milk out. They don't like the sound of the machine; it makes them nervous. We are still seeing how this works [chute structure], if it does not work well. I'd better use a fixed stall [mechanism that locks the cow's head in a trough, while it is milked], so that they go in with their heads.
>
> *(Fieldnote January 7, 2015)*

It was clear that this adaptation process was also felt by the cows, who at the beginning refused to be connected to the machines, while the ones who were connected reduced their milk production and had to be milked by hand after being milked by machine to express all of the milk. Otherwise, the cows would retain milk and develop mastitis. This change from a manual to a mechanical milking process represented a new intersubjectivity between the animal and the producer as well as an entire emerging corporality of man, machine, and cow.

In terms of innovation, MAGAP aspires to "diffuse" and "transfer" technology to producers with the aim of having the farmer and the animal immediately adapt to the technology. The result could be a quick increase in milk production. Nevertheless, the reality of absorbing this technological modernity involves a reorganization of the relations embedded in peasant farming practices that go beyond just an increase in productivity. Drawing on Arce and Long (2000), it can be argued that producers do not experience the arrival of technology as the disintegration of their old worlds. Instead they undergo reality as a living ensemble of imagined and felt experiences that may contrast, harmonize, and interrelate different materialities and agencies, embracing aspects of modernity and tradition. In the same vain, Bear and Holloway (2015), following Law and Mol (2008), go beyond the technological object itself, showing how a cow can be seen as subject of a series of overlapping "enactments" where the animal is not only a physical body, but also subject to infrastructure design, a factor of economics, a member of a wider group of cattle, and a living being with which a farmer may develop a close relationship. When a producer said, "The cows are stubborn today," this depicts how the cow performs a new corporality, involving a particular technology and its demands. This implies that cows are not only productive animals, but also individuals performing and expressing a subjectivity in relation to one another, the producers, and technology (Bear and Holloway, 2015).

María mentioned that in July 2011 an agreement between the Cattle Farmers Association of Santo Domingo (ASOGAN), MAGAP and the Municipality of Santo Domingo led to the Project for Technology Transfer for Development (TRATE 2011–2013), in which the municipality and ASOGAN bought seven cooling tanks to carry the insemination straws, and MAGAP committed to creating a linkage among producers' associations from seven parishes. In each association they trained one of the members to inseminate the livestock, while MAGAP would conduct regular gynecological check-ups on cows as well as hormone preparations (sync) and the person in charge from the association would identify the natural heat period and decide when to inseminate. Unfortunately, according to María, the program did not work because, "Some associations don't like to progress."

While a MAGAP technician sees the process of technology adaptation slowed by a peasant farmers' unwillingness to cooperate with technology adoption, it can be thought that although it was within the politically desirable targets of the nation-state, the policy was no longer achievable due to the instruments used by the nation-state. This bottleneck of policy implementation leads to a control dilemma in which the state mechanisms lose their effectiveness; while at the same time, peasant farmers' expectations of the state remain high (Beck and Lau, 2004).

Conclusion

Peasant farmers are able to use and develop material and social resources in many ways that reconstitute or transform localized situations, cultural boundaries, and knowledge (Arce and Long, 2000). This process occurs in a latent fashion and is in constant political tension with the nation-state, comprising development of political loyalties along with diverse forms of life. The appearance of non-state political actors and growing protest movements against neoliberal globalism provides support to a different local form of counter tendency to globalization (Beck, 2006).

In practice, Ecuador's official agronomic approach to rural development – food sovereignty – attempts to displace the knowledge and social organization of rural families. Ecuadorean policy is organized around notions of centralized administration, where the goals and means of transformation are conceived as the product of universal plans, thereby positioning the peasantry as an object vis-à-vis an authoritarian state. This act displaces rural experience, while undermining relationships and certain modes of existence and generating conflict.

Therefore, peasant farmers are re-adjusting concepts and categories that belong to the making of the modern nation-state or even rejecting them radically (Beck and Grande, 2004). The nation-state not only assumes the presence of a political society, but it also needs basic congruity between the scope of its competence to regulate social action and the range of social action itself. Both individualization processes and unpredictable socio-economic developments that arise within civil society and its organizations eventually break the nation-state's container (Beck, 1993; 1997), collapsing the presumed coherence between state-based regulation and the scope of both individual and societal action. This is a threat to the existence of political society, since it erodes the basic conditions for solidarity within society. Therefore, if the people of Ecuador want to ensure democratic self-determination and individual rights and liberties amidst rural development, they have to seek new social and political institutions adapted to this opening of the reality of alliances between people and things. The technocratic incongruity between the state's production aims and the materiality of rural livelihoods generates doubt over the possibility of state-led food sovereignty.

The incongruity between state policies and peasant farmers and farming challenge government regulations and technological packages, and it forces rural producers to be selective when adopting new technology. These assemblages lead to the emergence of a new intersubjectivity between producer, farm animals, and particular technology. These are new ways of active objectification (Bear and Holloway, 2015; Haraway, 1991) of "rural things" that follow the experience of the farmer and that open a window to thinking about technology and animal participation in policy, in this case for the purposes of food sovereignty.

In this chapter, I presented the producer, the milking machine, and the cows not as objects or as subjects but as a multiplicity of affect that generates a new corporality comprising each of the three. This new composition and how it operates is what the state cannot capture or understand when it objectivizes the peasant farmer and the process of production to a set of modernization standards and procedures. The elements of objectivization of the state do not match this new emergent corporality. Paradoxically, this increase in objectivity no longer produces an increase in consensus. Instead it leads to further encounters of dissent (Beck, Bonss, and Lau, 2003).

Hence, policies that aim to support peasant farming seldom align with the materialities of current practices. While the principles of food sovereignty hold the peasant farmer as the main actor for its fulfillment, the technocratic way in which the state applies food sovereignty remains disconnected from the peasant mode of farming (Van der Ploeg, 2009) and its rural materialities. The state's absence of reflexivity when assessing peasant-farmer practices makes food sovereignty policy detached from the matter of rural life, and as such unachievable. The intersubjective experiences that the peasant farmer goes through in confronting technology from the ministry depicts the need for more sensitive and sensible assessment of the multiplicity of rural life, which lies beyond modernization, progress and business-oriented agriculture.

Notes

1 Ecuador's Law of Food Sovereignty (LORSA), National Assembly, Official Registry No. 349, 27 October 2010.
2 Precinct is a political-administrative division of the fourth level with regard to a parish.
3 The scope of individual and societal action exceeds the boundaries of the nation-state that results from the welfare state and the subsequent individualization processes. This is seen as a basic experience of globalization (Braeckman, 2008).
4 In 2011 MAGAP started a national cattle improvement program through which the DPA of Santo Domingo received cooling tanks, semen injection equipment, and "improved" genetic material for cattle impregnation.
5 This local knowledge originates from experiential knowledge, rather than traditional knowledge. Brock, McGee, and Gaventa (2004) refer to the lived experience of poverty as being very disconnected from the poverty policy process, thereby holding experiential knowledge as less legitimate for the policy process.
6 The introduction of new technologies of automatization acts to remove the presence of humans in the farm. This leads to new spatio-temporalities of life in agriculture (Bear and Holloway, 2015).
7 The chute is a wooden or metal structure simulating a corridor in order to direct cattle from one pen to another, in this case it was used to tie the animals to it and keep them from moving during the milking process.
8 Tying the calf next to the cow can stimulate the animal to express its milk.

References

Arce, A. (1993). *Negotiating Agricultural Development: Entanglements of Bureaucrats and Rural Producers in Western Mexico*. Wageningen, the Netherlands: Wageningen Agricultural University.

Arce, A. and Long, N. (2000). Reconfiguring modernity and development from an anthropological perspective. In Arce, A. and Long, N., *Anthropology, Development and Modernities*, Abingdon: Routledge, 1–31.

Arce, A., Sherwood, S., and Paredes, M. (2015). Repositioning food sovereignty: between Ecuadorian nationalist and cosmopolitan politics. In A. Trauger (ed.), *Food Sovereignty in International Context: Discourse, Politics and Practice in Place*, Abingdon: Routledge, 125–142.

Bear, C. and Holloway, L. (2015). Country life: agricultural technologies and the emergence of new rural subjectivities. *Geography Compass*, 9(6), 303–315. doi:10.1111/gec3.12217

Beck, U. (1993). *Die Erfindung des politischen*. Frankfurt-am-Main: Suhrkamp.

Beck, U. (1997). *Was ist Globalisierung? Irrtümer des Globalismus Antwort auf Globalisierung.* Frankfurt am Main: Suhrkamp.

Beck, U. (2006). *Cosmopolitan Vision*. Cambridge: Polity.

Beck, U. and Grande, E. (2004). *Das kosmopolitische Europa. Gesellschaft und Politik in der Zweiten moderne*. Frankfurt-am-Main: Suhrkamp.

Beck, U. and Lau, C. (2004). *Entgrenzung und Entscheidung: was ist neu an der Theorie reflexiver Modernisierung?* Frankfurt am Main: Suhrkamp.

Beck, U., Bonss, W., and Lau, C. (2003). The theory of reflexive modernization: problematic, hypotheses and research programme. *Theory, Culture & Society*, 20(2), 1–33.

Braeckman, A. (2008). Reflexive modernization and the end of the nation state. On the eclipse of the political in Ulrich Beck's cosmopolitanism. *Ethical Perspectives*, 15(3), 343–367.

Brock, K., McGee, R., and Gaventa, J. (2004). *Unpacking Policy: Actors, Knowledge, and Spaces in Poverty Reduction in Uganda and Nigeria.* Kampala: Fountain Publishers, 1–26.
COPISA (2014). Quienes somos: Conferencia Plurinacional e Intercultural de Soberanía Alimentaria. Retrieved from: www.soberaniaalimentaria.gob.ec/?page_id=1863.
Deleuze, G. and Guattari, F. (1988). *A Thousand Plateaus: Capitalism and Schizophrenia.* London: Athlone Press.
Haraway, D. (1991). *Simians, Cyborgs, and Women.* New York: Routledge.
Law, J. and Mol, A. (2008). The actor-enacted: Cumbrian sheep in 2001. *Material Agency.* New York: Springer, 57–77.
Ploeg, J.D. van der (2009). *The New Peasantries: Struggles for Autonomy and Sustainability in an Era of Empire and Globalization.* London: Earthscan.
Stenner, P. (2008). A.N. Whitehead and subjectivity. *Subjectivity*, 22(1), 90–109.
Whitehead, A. N. (1933/1935). *Adventures in Ideas.* London: Cambridge University Press.

7

FORGIVE ME FOR BEING HUMAN

Wirikuta nomadism and rebellious Peyote

Oscar F. Reyna

Introduction

Our common sense tells us there is a direct correlation between *food*, as the matter-input for feeding body organisms, and *life* – understood here as the consequence of an impermanent set of vital movements. Still, the general asseveration that "food is necessary for life," in whichever political position it actualizes, must be critically observed, for such concepts will vary depending on an actor's position in a world "affected by contemporary science and biotechnologies affecting the very fiber and structures of the living, and altering dramatically our understanding of what counts as the basic frame of reference for the human today" (Braidotti, 2013: 40).

In this chapter I pose some questions on globalized, standardized, and taken-for-granted notions on food, life, and the restlessness that occurs when discussing the present and futures of such topics. This chapter will not get into debates on access–labor, scrutinization–sanitization, and productivity–entrepreneurship on food and its capital-driven linkages present in the processes of production, circulation, and consumption – all framed within the "objective" ways the world functions today. Conversely, I will explore a critical reading of such debates by looking at forgotten affective practices related to a series of "others": kinds of food (food for the soul), kinds of agriculture (technologies of non-cultivated food), ways of being nurtured (socialization by eating it), and the establishment of relations around such food practices (economic relations in simultaneous pursuit of collective well-being and body–mind decolonizing practices). I frame the latter under a broad notion of intersubjectivity, understood as the practice of relating to others,[1] "including nonhuman and *Earth* others, resting on an enlarged sense of interconnection" (Braidotti, 2013: 48).

Discussion will be guided by the inter-ontological practices of eating Peyote in the Wirikuta desert in northern Mexico, presently threatened by the arrival of extractive projects. An attempt to symmetrically understand the entanglements of humans and more-than-humans in this framework (Latour, 2005) will be addressed by following an interspecies approach (Kirksey and Helmreich, 2010). In this chapter, Peyote (the food for the soul) is seen as co-creator of rebellious acts, whether as affective-motors in the cosmopolitical[2] struggle for conserving the Wirikuta Sacred-Natural Site or favoring the simultaneous awakening of creativity and critique[3] by motivating the exploration of other ways of eating, moving, consuming, and living. The consumption of the "sacred food" will be set as a turning and de-territorializing[4] point in which pre-existent categorizations of the "objective" world become fluid and dissolve, simultaneously dissolving "objective" notions and practices occurring in it. The central argument is that eating Peyote in Wirikuta is a multispecies act that situates the cactus as a companion species (Haraway, 2003) whose genetic properties trigger affections that have invited those humans interested in preserving the area from extraction to become "nomads." This process occurs in multiple levels of movement that start when actors are looking for the cactus and consume it. It finds deep significance when they finally get involved with the situated community to perform "affirmative politics," combining critique and creativity and an engagement with the power of life itself (Braidotti, 2012). In the process, actors seize the opportunity of globalizing social relations in "progressive ways" (Massey, 1994, 2004). They think about and enact in the social space in processual and open-ended forms, they become aware of inner conflicts but do not get stuck in them, and, instead, they direct efforts towards emphasizing the situated uniqueness of life that is all inclusive.

In this analysis, the nature/direction of social action conveys the implicit idea that alliances of human, more-than-human actors and the emergent ways of the production of subjectivity are deeply interlinked. Without understanding such world encounters, no anthropological approach can be broad enough as to bring down the dividing walls of asymmetry (Latour, 2005). I consider actors in Wirikuta to be the mobile/mutable/multispecies centers around which diverse movements, territorialization practices, and social actions are created. I will use "movement" as a central concept: the social actions undertaken in different worlds, which spin in and around the practice of eating Peyote among non-indigenous urban people who visit Wirikuta.

Peyote: a resistant, disobedient cactus

The sacred desert landscape of Wirikuta, and the assemblage (Deleuze and Guattari, 2011) of worlds it gives life to, are currently jeopardized by modernization ventures. Two large-scale transnational mining projects aiming to extract gold and silver were stopped by two moratoriums achieved by activists in 2012 and 2013 respectively (Enciso, 2013). Nevertheless, threats to the region have been recently maximized with plans for building a dam in La Maroma, the expansion of tomato agro-industries in San Cristóbal, and the construction of a toxic waste dump in Rancho Palula (Chávez, Chávez, and Gavilán, 2015). Should these projects take place, they would entail severe environmental damage, that is, landscape disruption, water, soil and air pollution, eradication of endangered species, disablement of soil in traditional rural areas, and an increase in regional water stress. Such scenarios have provoked the emergence of a cosmopolitical movement constituted by heterogeneous actors (see Reyna and Arce, 2015; Lamberti, 2014; Liffman, 2012), one of which is the Peyote cactus, endemic to the region. This movement converges around the protection of Peyote from extractive industries and as a consequence, Peyote has become an icon of struggle. State representatives and investors have a hard time accepting its presence because of the connections folded in and around its body. It has the capacity to create networks whose mobilizations challenge the legitimacy of extractive and infrastructure projects, while also forcing crucial debates on the rights and legitimacies of the existence of other ways of worlding (cf. Blaser, 2012), disregarding these networks if they are tagged as indigenous, peasants, small-scale miners, hippies, resources, species, and so on. Peyote, it could be argued, has enabled the resistance to development interventions that threaten its own existence and the realities of others connected to it.

But resistance is indeed inherent to the nature of Peyote. Growing slowly with optimal use of energy and water, being able to survive despite long periods of drought, and having to face the warmest days followed by the desert's coldest nights, it is indeed an embodiment of resistance (Figure 7.1). Peyote species inhabit a wide expanse of the desert, disregarding the imaginary borders imposed by the seven municipal demarcations that make up the region. They disrespect state boundaries as they do private and communal property. Sometimes their heads are even spotted inside greenhouses in the United States, discretely sharing the soil irrigated with underground water to grow export quality tomatoes. Thanks to their reproductive abilities, their seeds can go through wire fences, which are recently appearing to divide boundaries of former *Ejidos*; they trespass and grow in the middle of either abandoned or active plantations. For mestizo peasants, Peyote is a cactus as inoffensive as any other. If they find some within their lands they usually let them grow. However, they are aware that Peyote may be of interest to someone else, in which case they commonly accept people coming into their land to take them. In contrast, the cactus is highly valued among the Huichol ethnic group. Every year, designated Huichols are entrusted to undertake a ritual pilgrimage to Wirikuta to collect Peyote, which is essential for their ceremonies held during the year. It is a common sight to find Huichols walking in the middle of the desert, visible because of their eye-catching costumes in white and bright colors. Sometimes, non-Huichol people accompany them to get closer to the Huichol worldviews.

While Mexican legislation allows indigenous peoples to use Peyote for ceremonial purposes, it forbids and criminalizes its use by non-indigenous. However, its use is still common among non-indigenous willing to experience the psychoactive properties with varied purposes, ranging from merely recreation to involvement in neo-shamanic courses while undertaking spiritual tourism (Bassett, 2011) and/or carrying out deeper quests for meaningful answers in response to modernity crises in all sorts of re-Indianization processes (Liffman, 2012). As Peyote contains 60 alkaloids (Batis and Rojas, 2002), humans eat it as a way to connect with their inner self, expecting to experience healing processes and/or psychological changes. An interesting contrast emerges when noticing Peyote's disrespectful behavior towards the law. Police and militaries have created problems for those trying Peyote; superficially categorizing them in terms of race and right to access it, with frequent misunderstandings. However, the cactus does not stop attracting people indiscriminately. Class, gender, ethnic, national and religious qualifications are secondary when it comes to finding it and consuming it. Peyote is, once again, disobedient.

Kauyumarie: Peyote's elusive character

Practices of finding and eating Peyote in Wirikuta entail significant processes of movement. Since consuming it always involves the risk of becoming disobedient as well, a strong will is required. Disobedience and subjective movement are related in this chapter to becoming a nomad, to be able to move from assumptions about the world one lives in, or from the comfort that ideas about food, homes, jobs, positions and relations give to contemporary Western or Westernized humans. Movements happen when eating Peyote, but they also happen when trying to find it. Walking is an important process, which creates a specific sense of place that is reaffirmed by consuming the plant. This starts by moving in the middle of the desert to hunt food for the soul. It encompasses practicing the art of walking-observing like the Chichimeca people did in Wirikuta before the Conquest. Becoming nomad is a human ability that is being forgotten in contemporary sedentary positions, very common in urban concentrations. Nomadism then is simultaneously corporeal and mental–spiritual. In this sense, the practice of eating Peyote involves, just like the defense of Wirikuta, centrifugal and centripetal forces creating connections both inside the self – reflections, existence – but also mobilizing the encountering bodies by walking and socializing, establishing networks with the otherness to create new possibilities of making things happen.

Those who have undertaken the quest to find Peyote know that a certain ability to endure long walks under the blazing sun is essential. Patience and orientation are needed and tested when finding oneself in the middle of nowhere, surrounded by endless monotonous and labyrinthine bush patterns spread in the greyish sandy ground. It is essential to be aware that rattlesnakes, coyotes, or poisonous thorns own the pathways; it is their homes that humans walk into with the aim of getting the benefits of mescaline, the main alkaloid contained in Peyote bodies. Hunting the sacred cactus requires a high sense of alertness, but also sensitivity.

Mestizo peasants of the region recall stories about foreigners trying to find it unsuccessfully, mocking those who have looked for it behind leaves of bushes, digging deep holes in the ground, or climbing on top of spiny Palmas Chinas to have a better view. Tragic stories are important warnings, recalling people who passed away while looking for it, their corpses found kilometers from the point they started walking. It is recognized that the quest is always a matter of life or death.

Failures to find Peyote are explained by Huichols in specific ways. They argue that those people came to the sacred desert ignoring the Peyote's sacred character, unaware of its sneaky behavior. Huichols compare it to "an elusive deer or a mischievous child" because it is well known that "it likes to hide." In fact, Kauyumarie, one of the Huichol designations for Deer-Peyote literally means "astute" and is associated with its ability to dodge the wiles of hunters (Diguet, 2005: 174). It is commonly said that cactuses are the ones who pick the persons that can – or cannot – eat their fleshy bodies. When being close, Peyote plants tempt a person's character while watching them, and after some trials they decide to appear or remain invisible to the eyes of the unprepared. Their reserved personality resonates with the ecology of the cactuses. Despite usually being found grouped in "families," often growing under *gobernadora* bushes, it is hard to spot them because they are usually camouflaged with the stony-dusty ground around them. The latter may be a form of self-defense against predators since it lacks thorns unlike most of the cactuses in the Chihuahua desert. It is also common to find them buried, completely covered by the sandy soil. It is said they bury themselves following a self-protective instinct in dry times when the risk of dehydration is high. Such behavior actually supports the Huichols' strong statements on collecting Peyote: "You don't harvest Hikuli,[5] you just cannot; you have to hunt it."

FIGURE 7.1 Food for the soul: the Peyote cactus

Source: Photo courtesy of Edith de la Rosa, 2014.

Besides knowing the multiple mythological explanations, which tend to vary widely, of Peyote and the world, for Huichols, sacredness implies being able to establish a particular relation with the self, the more-than-human specificities of the territory, and the agencies and ontologies of species (whatever their taxonomy) that are interwoven with the sacred land. This is well known to non-indigenous people who decide to get acquainted with the cactus through the Huichols' views, creating new mixed identities through contact as a result of the mutual affections of living with, following and learning from them. The cosmopolitical movement for protecting Wirikuta from the threats of extractive industries have made Huichols, Peyote, and Wirikuta very visible worldwide. As a result, a wide array of people interested in Peyote, ecologist practices, alternative medicine, psychedelia, social movements and indigenous worldviews, come to Wirikuta to experience other ways of enacting with the world and other ways of experiencing themselves. Eating Peyote is the departure and turning point.

Self-sacrifice for the sake of reproduction

For Huichols, Peyote is identified as the living substance (the heart/flesh) of Tamatsie Kauyumarie, the Huichol's most important mythical hero. According to narrations, it was a primal divine deer who after a long journey of escape and persecution decided to offer itself to the first hunters as an act of compassion. This self-sacrifice happened in Wirikuta simultaneously with the very first sunrise, leading to the establishment of time and space, as we currently know it (Neurath, 2000: 73), and to the advent of the first deities of Wixarika's pantheon (Gutiérrez, 2010: 193). Since Hikuili embodies sacred qualities proper to ancestors/deities, special procedures involving a complex succession of rituals are followed. These involve a public confession of personal sins to avoid mental disturbances in the trances. Fear or bad feelings have to be left aside, otherwise people expose themselves to being "punished" by the plant causing bad sensations. It is also important to make an offering to it; it may be a coin, a calabash, or even bigger offerings entailing complex and difficult physical tasks. When the time arrives to cut the Peyote, a golden rule must be followed indisputably: only the "shell" (the apex) can be cut. The root must remain untouched. By doing this, the plant will grow again: "Hikulis know how to thank." Such procedures and kinds of awareness are indeed, another kind of agricultural technology.

The reproductive biology of Peyote is amazing. When cutting only the apex and leaving the root untouched, as the Huichols recommend, it grows again in different ways. Ivan Korzeniowski, an amateur botanist and Peyote lover, explained it in this way: "If the plant is cut in half then the inner part of the apex can remain, so the plant continues to grow normally from its apex. But if the plant is cut further down then the pinnacle of internal growth would be cut, so the plant would birth side heads, instead of continuing to grow from its center" (Figure 7.2).

Ivan's explanations refer to two ways of surviving: in the former, Peyote regenerates, while in the latter, it multiplies. This explains another attribute the

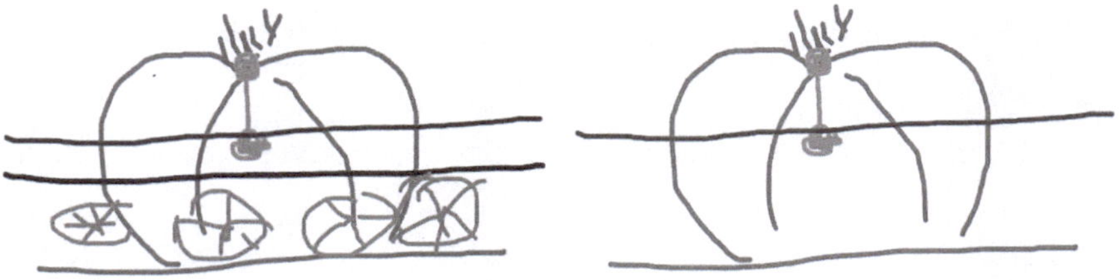

FIGURE 7.2 Tillering (left), regeneration (right)

Source: Drawings courtesy of Ivan Korzeniowski, 2014.

Huichols give it: "Peyotes like to be cut," "They enjoy it," "They need to be cut," "It is not bad to cut them," they usually say when being challenged by conservationists or people concerned about damaging their wild status. Peyote sacrifices itself for others to survive, but when doing this not only does it survive, it also multiplies. The following ethnography explains how Peyote is not only a rhizome in shape; it enables rhizomic social configurations around it.

Going to the encounter: the bio-construction workshop

In February 2014, a bus with 40 people arrived from Mexico City at La Flor del Desierto eco-culturistic facility. They came specifically to learn about sustainable construction techniques guided by Javier Bonifaz, a renowned Peruvian artist/bio-constructor. Organi-K (an urban ecologist organization based in Mexico City) and Colectivo Patas Verdes (an association of local people and newcomers currently living in Ejido Las Margaritas) organized the workshop with the help of people of Refugio de Coronados. The first two groups arranged logistics, publicity, and transportation from the city and the people of Refugio de Coronados helped with the hosting, preparation of food, and facilitating local resources, which was remarkable considering the isolation of the location and water scarcity in the area.

During the four days of the encounter, participants with urban and middle-class backgrounds became engaged with local people. One of the aims was to improve the material conditions of La Flor. Outdoors, participants took turns mixing mud barefoot to create adobe. Others worked closely with peasants from the community in building an outdoor mud oven. Villagers were interested in learning the *Nubian Vault Technique* taught by the instructor. They wanted to improve their traditional ovens in terms of resistance and performance, so they participated actively and hand-in-hand with the visitors (Figure 7.3).

Indoors, a different group sculpted and colored four animal-shaped relief murals depicting an eagle, a deer, a hummingbird, and a puma. The drawings were designed by Javier Bonifaz to give the inner walls of La Flor del Desierto an artistic imprint. Nevertheless, participants attributed particular meanings to them. Deva, a

FIGURE 7.3 Neighbors (locals and newcomers) and visitors, together, building an open-air mud oven with collaborative work in La Flor

Source: Photo courtesy of Edith de la Rosa, 2014.

holistic therapist interested in sustainable ways of living and women's empowerment, strongly affirmed that they were created to protect the facility: "*La Flor* needs protection and strength to continue with its labor as an alternative place for promotion and protection for Wirikuta against *las mineras*."

On the second day of work, people continued with and expanded their chores. By midday, some persons came into the kitchen to help women prepare vegetarian dishes for dinner. Food was made with local products like goat cheese, *nopales*, beans, tomatoes, and of course, maize tortillas, all provided by the people of Refugio. Their products were also offered for sale in a small improvised stand that was put together: organic tobacco, Huichol handicrafts and balms, conserves, and shampoos produced by a Women's Cooperative from Ejido Las Margaritas (Figure 7.4). A small handwritten cardboard sign encouraged participants to be part of local, alternative and fair consumption by buying some of the products on display.

By this time of the day, the assistants noticed a group of three persons were missing. The absence of one of them was notable because he had actively been fetching water the day before. Someone said they had seen the group walking into the desert. Since they were seen accompanied by one of the organizers as a guide, the rest continued doing their chores normally.

Following the call

Right after putting the last layer of mud in the top of the ceramic oven, David felt a call:

> I knew of its existence, I knew we were going to a sacred place on this volunteer trip. My first intention when coming was not to eat it. It is hard to describe this call, like a thread from the heart pulling me and saying come, come. Of course I was informed that it is classified as a dangerous addictive plant. People are trying to change this status from class 4 to class 3, as far as I remember. The problem is the result of an ancient history going back to the colonial period. This happened because they realized its healing power. The fact that it is forbidden nowadays is just irrational. We are collateral victims of the war against drugs. We are being judged for eating a plant that is medicinal, that heals. Peyote is food for the soul. But this generates confusion about it all related to Wirikuta and its defense.

Because of this information, David remembered having an internal debate about hunting for Peyote. Having an encounter with Peyote would distract him from the main objective of the workshop, helping local people. However, he and his colleagues decided to answer the call. After having expressed interest to one of the people in command of the community, who understood this was a special request, they embarked on their walk:

> It was a nice walk for self-knowing ourselves. When walking, I could not see any of them. It was getting dark, but just before sunset I saw it, I had found it! Our guide suggested using the Wixarika ways of Peyote hunting. That was the key to find it, actually. First, each one of us spoke to and presented ourselves to Wirikuta. Then, we all explicitly asked for permission to enter to the desert. When finding it, we had to present ourselves before the Hikuli. Then we asked for permission to commune with it. We left a modest offer on top of the cactus. By that time, we could already see other Peyotes around. We ate some of them, slowly. The first bite was shocking; its sourness, the first thing I sensed, was followed by my body shaking. Later on, I could feel its strength, its energy. And then I could see things more vividly, but not only with my eyes, but with all senses. I could feel skin deep.

After having eaten some Peyotes, they returned to the camp. But then they found themselves with a bunch of the cactuses they had cut. Since these cactuses became visible, they interpreted it as a gift, and therefore felt that the Peyote plants needed to be collected and shared with workshop participants.

Sharing the desert's gift

By nightfall, the missing group reappeared with the idea of making a bonfire. It was lit in the center of one of the spiral shaped stone altars in the desert courtyard. As the fire was burning, some participants who were still awake joined the circle. They carried out collective dances and chants, typically performed in New Age Mexican ceremonies. People kept the fire burning until the first sunrays appeared.

The next morning, the sound of people preparing to walk into the desert could be heard from the tents. Some people woke up and joined instinctively while others preferred to stay asleep. The ones who woke up followed a guide for about 40 minutes in a silent Indian file until stopping at a random spot. They formed a circle.

The guides collected dry bushes to make a small fire in the center. One of them distributed tiny wax candles. A box with colored ribbons was passed around for people to choose whichever ribbon they wanted. An explanation was given: "Vas a tejer tu rezo, tu intención en la velita, luego la vas a ofrecer" (You will knit your prayer, your intention in the candle, then you will offer it). They all created different ribbon patterns in our candles silently. Just before the moment of offering, David stood up, and looking directly to the sun he offered "his work" to Grandpa Fire.

Then he presented a red *paliacate* containing slices of Peyote cactus he had collected the day before. Peyotes were circulated among assistants with the help of one of the organizers. While sharing, some persons were given the Peyote slices individually by the organizers in a micro-ceremony that resembled the Catholic communion. Initially, some participants hesitated eating their slice of Peyote, but at the end everybody did.

At that point they all started taking turns randomly going to the fire to light their candles. Each person pronounced prayers, chants, and petitions. A young couple gave thanks to the Deer God and Wirikuta for conceiving a new soul: they were expecting a baby. Some people made emotive calls for the guardians/ancestors of the place to protect Wirkuta from the threats of mining companies. A teenager, who seemed to have strong ties with the Nahua ethnic group, brought a decorated maize cob, and while offering it he made a call for "respecting indigenous traditions and humble ways of living" like the ones of his own grandparents. After every single person said a prayer, they were asked to walk clockwise around the fire and then follow the guide back in a silent and reflective Indian file.

Touching deep fibers

David was born in Mexico City and soon will be moving to Chalmita, in the mountains. His parents are Mexican and he has some Swiss ancestry. He started studying industrial engineering and computing, but he had to quit because of an existential crisis that led him to question his entire being. As a result, he spent four years in Europe traveling while trying to answer the question of what he wanted in his life and what he did not.

In his own words, he remembers being blinded by a traditional education scheme and deeply immersed in a severe pattern of consumption. When he realized this he decided to change his life. He started by reforming his own understanding of himself as a human being. Then, he took a course in biodynamic agriculture in Switzerland and afterwards, he felt an impulse to come back to Mexico, because he "needed mangos and avocados again." Then the biggest cultural shock ever happened to him. He found himself confronted with what he had left before: "Family, friends, previous relationships and basically the world of consumerism. The destruction and abuse we live in, shocked me. As this was painful I decided to move myself again." His concerns led him to San José del Pacífico, in Oaxaca, where he had his first psychedelic experience. By walking into the forest he could find some mushrooms. His guide oriented him on how he should consume them. After eating them, sensations and messages came.

> While I was there I could hear the chainsaws of persons cutting down some trees. Since mushrooms give you very keen senses, I realized I did not imagine the noises; they were really there. The sound became more and more intense until it started to hurt me. Hearing that happening just a few meters from where I was resting actually hurt me. Just thinking of how those giant wonderful beings were being chopped brought me to tears. I ranted on about why they were doing this. Can they not see what they were doing? I asked myself. But then … then I realized it was myself who was chopping those trees down. It was I with my own lifestyle who was encouraging it. I started to think about my own house where we have wooden doors. We have a wooden table that must have come from some tree. I had to stop this, at least from and for me, but then I realized everything I should let go. Then I started walking into the forest, following an indescribable impulse for life that into my mind was screaming to myself: "I don't want to kill anyone else." Suddenly, I realized that every step I was making I was stepping on something, I was killing that something. From that impression I stayed completely static, with no will to step further. Then the real change came. Fully conscious, I decided not to stop walking anymore. Instead, every step from now on would be a step to promote life, to take care of it. Then, a phrase came to me: *Please forgive me for being human.*

Such an experience led David to explore other ways of living, even the basic human need of eating. He followed a "Pranic Process" which involved fasting and living off just water and "Prana," the energy of the sun. Although it was difficult, he became aware that a radical detachment from consumerism was possible: "Consumption is the key. We really do not need anything to live. But I do not want to need anything, and then I need to start to accept it all. And by the time I enter into a conflict, I fight with myself. That takes energy, therefore I start to need more things."

David is constantly trying to consume less. He tries to take every action more consciously to save energy and materials. Another starting point he has identified is to dissolve segregation. He explains, "Sometimes we want to get all the bugs away from our houses, to feel isolated. The challenge is to learn how to integrate." For David, the way of doing that can start with actively engaging in permaculture and conscious design to reduce the impact of human activities.

During our interview, David mentioned *value* as another important element, emphasizing, "The value of life, the real value of existence." A reaffirmation of how to live and operate came with Hikuri, in Wirikuta, where an external force offered him "the direction and impulse on how to undertake a struggle for respecting and caring for the earth, starting with care for oneself and for others."

A further impact of his visit to the desert was to get engaged with the community he met. As a matter of fact, the day in which participants were coming back to Mexico City, a sort of improvised movement happened among the organizers who were in charge of the selling stand. David had bought five boxes with an assortment of the products of La Flor del Desierto. His plan was to commercialize them in eco-touristic villages in Cancún with which he was associated. Since there was not a large stock of products, the organizers had to collect products from different points in the desert to secure the transaction, which represented a huge opportunity to mobilize the products for sale. Besides, he continued to support volunteer trips in different modalities like the bio-construction workshop. Civic organizations and local people get economic benefits. Affection among them and with the land is created. Knowledge mobilization is also evident, and building with eco-techniques has gradually changed people's life in the desert.

Years ago it all started with Lalo, a single newcomer aiming to teach women how to make balm, preserves, and other products using wild medicinal and edible plants. Later on, a woman's cooperative emerged: La Flor del Desierto, which has become somewhat independent because earning profit and diversifying circulation networks for their products has created fissures in the local social fabric. But diversification of interventions in communities has brought more experimental benefits.

Clear examples of benefits are the planting of a fruit tree orchard that is fertilized with compost, irrigating with rainwater in the dry complex of La Flor, and community work focused on dredging old irrigation canals in Ejido La Cañada. "Tillering" of participants in the social defense of Wirikuta is underway. Local people have gotten access to new trade routes for their own products like native avocados, quince, tortillas, and embroidery while city people get the opportunity to become nomad, even for a weekend, and establish relevant networks for strengthening the resistance.

An obscure side effect of these actions is, ironically, that Wirikuta is now commonly known as a place of power. Such connotation directly links the land with the sacred Peyote cactus, which is on one hand an attraction, but on the other hand a burden. This generates problems for activists because, in spite of not denying their links with Peyote and framing them within their experiences with Huichol people, it becomes a stigma when stating their position against development of the region.

FIGURE 7.4 Assortment of locally produced products for sale in the workshop

Source: Photo courtesy of Edith de la Rosa, 2014.

People in favor of mining companies constantly argue that activists "just come to exploit the cactus and to get high." A counter argument often concerns decriminalization of Peyote and its recognition as "a medicinal plant that should be available for all of mankind." In the bio-construction workshop most people were builders, architects, engineers and persons interested in permaculture construction; not all were aware of the existence of Peyote, which means not all of what is said by mining companies' sympathizers is true. However, it is a fact that affective experiences related to Peyote actually help interweave especial ties that are translated into the broadening of networks, commercialization, and labor in exchange for the invaluable experience of sharing Hikuli.

Nomadism, subjectivity, and engagement

Peyote's quality as sacred food makes it impossible to commodify (Kopytoff, 1986), which is reaffirmed by its status of a wild plant. "It is not cultivated, nor the product of my labor," reflects Tsing (2012) when following the delight of matsutake mushrooms. Peyote likes to be cut; it wants to grow, reproduce, and multiply. Humans stimulate this nature when dissecting them. Peyote will keep on multiplying regardless of any state-capital plan.

The movement to defend Wirikuta will keep on resisting as well, since their affective experiences trespass boundaries, such as in the case of Peyote seed. Huichol traditions, European quests, American adventures, and Mexican radical decolonization of people's minds will keep on happening in Wirikuta. Acceptance and diversification, implicit in the Peyote body will keep on producing subject - ivities in line with their actions, impacts, and potentialities. This is an embodied and embedded production of subjectivity (Braidotti, 2012), achieved by the decision to become a nomad, or similarly by being in charge of permaculture, seeds, plants, food and consumption. By following Peyote, its affections, and its influence in producing new subjectivities and intersubjectivities in resonant resistance and solidarity, other kinds of values emerge and with them new questions arise. What do we eat and what for? What do we feed others and how?

Becoming a nomad-of-the-body in order to eat an uncultivated food that enables one to become a nomad-of-the-mind is an account of the posthuman, in which the objective world becomes fluid, and with it, humans enter into a constant state of quest and reflection. Along the way, values of profit, democracy, autonomy and equality are progressively substituted by values of well-being, solidarity, co-creation and exceptionality.

Fragility however, is a fact concerning such emergent configurations. This reflective exercise on exploring food (its kind and our kind) as important as this might be, is one giant leap towards the material grounding of the world's situated claims for new ethical values (Braidotti, 2012: 49). On such a quest it is important to pose the question: is the use of Peyote required for everyone to be aware of human and nonhuman interactions and the values it entails?

It is probably a pending task to find out in which other aspects of everyday life and with what kind of others we, as humans and nonhumans, can find such rebellious, resistant, self-sacrificial inspiration to challenge the status quo of excesses inherent in capitalist consumption. We still need to explore beyond our borders. Becoming nomad, regardless of the pathway, is now more than ever imperative.

Notes

1 The human–object relations addressed in this chapter will be seen through a post-humanist light that allows focusing on an open critique to self-centered individualism and its assumed practices–processes around the body and the social actions of eating, which constitute a contrasting empirical approach to alternative perspectives provided in this volume.

2 The struggle for Wirikuta is seen here as cosmopolitical since "it has no representative, no one talks in [its] name" (Stengers, 2005: 1003), see Reyna and Arce (2015).

3 "Creativity and critique proceed together in the quest for affirmative alternatives which rest on a non-linear vision of memory as imagination, creation as becoming" (Braidotti, 2013: 165).

4 Here, deterritorialization is understood in Deleuze and Guattari (2011: 156) as the decontextualization of any set of relations.

5 Hikuli is the Peyote's Huichol name.

References

Bassett, V. (2011). *Du tourisme au neochamanisme*. Paris: Harmattan.

Batis, A. and Rojas, M. (2002). El peyote y otros cactos alucinógenos de México. *Conabio. Biodiversitas*, 40, 12–17.

Blaser, M. (2012). Ontology and indigeneity: on the political ontology of heterogeneous assemblages. *Cultural Geographies*, 21(1), 49–58. London: Sage.

Braidotti, R. (2012). *Nomadic Theory. The Portable Rosi Braidotti.* New York: Columbia University Press.

Braidotti, R. (2013). *The Posthuman.* Cambridge: Polity.

Chávez, C., Chávez, T., and Gavilán, I. (2015). *La victoria y el ojo de venado; el tiradero tóxico de las mineras en Wirikuta.* 13 diciembre. Frente en Defensa de Wirikuta, available at: www.frenteendefensadewirikuta.org/?p=6639

Deleuze, G. and Guattari, F. (2011). *A Thousand Plateaus. Capitalism and Schizophrenia.* Minneapolis, MN: The University of Minnesota Press.

Diguet, L. (2005). El idioma huichol. Contribución al estudio de las lenguas mexicanas (1911). *Por tierras occidentales: entre sierras y barrancas.* México D.F.: Centro de estudios mexicanos y centroamericanos, 497.

Enciso, L. A. (2013). Suspenden todas las concesiones mineras que afectan la región sagrada de Wirikuta. *Periódico La Jornada.* Viernes 13 de septiembre, 38.

Gutiérrez, A. (2010). *Las danzas del padre sol: ritualidad y procesos narrativos en un Pueblo del occidente mexicano.* Mexico D.F.: Universidad Nacional Autónoma de México.

Haraway, D. (2003). *The Companion Species Manifesto: Dogs, People, and Significant Otherness.* Chicago: Prickly Paradigm Press.

Kirksey, S.E. and Helmreich, S. (2010). The emergence of multispecies ethnography. *Cultural Anthropology*, 25, 545–576.

Kopytoff, I. (1986). *The Cultural Biography of Things: Commoditization as Process.* Cambridge: Cambridge University Press.

Lamberti, M.J. (2014). ¿El Avatar mexicano? Una sociología no sociocéntrica de los conflictos derivados de la instalación de empresas mineras en territorios indígenas. *Intersticios Sociales.* El colegio de Jalisco, Vol. 8.

Latour, B. (2005). *Reassembling the Social. An Introduction to Actor-Network-Theory.* Oxford: Oxford University Press.

Liffman, P. (2012). *El movimiento de lo sagrado por Wirikuta: el límite de lo oculto en la cosmopolítica Wixarika.* Mexico D.F.: Ediciones Museo Nacional de Antropología.

Massey, D. (1994). *A Global Sense of Place. From Space, Place and Gender.* Minneapolis, MN: University of Minnesota Press.

Massey, D. (2004). Geographies of responsibility. *Geografiska Annaler: Series B, Human Geography*, 86, 5–18.

Neurath, J. (2000). El don de ver. El proceso de iniciación y sus implicaciones para la cosmovisión huichola. *Desacatos* 5. Centro de Investigaciones y Estudios Superiores en Antropología Social.

Reyna, O.F. and Arce, A. (2015). Cosmopolítica, patrimonio y contradesarrollo. La modificación de espacios de acción actoral ante el riesgo de la explotación minera en el sitio sagrado natural de Wirikuta. In M. Gámez (ed.), *Minería y capital transnacional sobre un territorio en riesgo. Análisis interdisciplinario sobre el sitio sagrado natural de Wirikuta.* San Luis de Potosí, Mexico: El Colegio de San Luis.

Stengers, I. (2005). The cosmopolitical proposal. In B. Latour and P. Weibel (eds), *Making Things Public: Atmospheres of Democracy.* Cambridge, MA: MIT Press, 994–1004.

Tsing, A. (2012). Unruly edges: mushrooms as companion species. *Environmental Humanities*, 1: 141–154.

8

FEEDING PARADISE?

Corporeal food citizenship in the Galapagos

Christine Franke, Jessica Duncan, and Stephen Sherwood

Introduction

The Galapagos Archipelago draws its romantic legacy from the writings of Charles Darwin, who, by recognizing the unique properties of the islands, challenged the science and religion of his day. Today, the widely held perception of the Galapagos is that of the birthplace of evolution: a "natural laboratory" set apart from space and time by its pristine and unique qualities. The Galapagos is subject to a popular idealism that promotes the conservation of the islands' raw nature, above all else.

Underlying this environmental conservationism are relational representations of appropriate coherency between people and the environment. Masking the "intensely disputed renderings of socio-natural relationships" (Wolford, Lu, and Valdivia, 2013: 93), such representations promulgate economic and political relations that separate "nature" from the "social." A popular construction of the Galapagos as a pristine ecosystem to be enjoyed by humans, but not to evolve with humans, raises serious questions about the limitations of environmentalism in general, and conservationism more specifically. As Quiroga (2009) explains, the Galapagos of our collective imagination is the result of a social construction that has served to reify the islands as a place for studying evolution and not a place for humans to live and be.

Over the last decade the Galapagos has experienced exponential growth in tourism, with the Galapagos National Park reporting 205,000 visitors in 2013, nearly ten times the permanent population of the islands. The population of the Galapagos has more than doubled from about 10,000 inhabitants in 1990 to more than 25,000 in 2010. Concurrent with tourism, the economic growth rate of the Galapagos since 1999 has been nearly ten percent. Provided the high migration elasticity of the islands, Taylor *et al.* (2006) argue that for every US$ 3,000 increase in income, another person moves to the Galapagos from the mainland, contributing to population growth, infrastructure development, energy demand, and overall consumption and waste.

This situation introduces a bind: an inevitable conflict between the Galapagos as a site of untouched nature and the Galapagos as a site of economic activity, growth, and livelihood. Economic dependence on tourism presents a second bind. Tourism to the islands is broadly acknowledged as pioneering "ecotourism" (Quiroga, 2009), seeking to be environmentally sound, small-scale, and to enrich the lives of local residents. Yet, as Brewington (2013: 106) warns, "The sudden growth and expansion of the [ecotourism] industry in the Galapagos has transformed this economic activity into a threat to conservation and social practices." The precipitous growth in ecotourism to the islands has increased the demand for food to feed the local population, those who bring the tourists, as well as the tourists themselves.

The social-economic and ecological schisms resulting from the unequal distribution of benefits from tourism place Galapagos' society in a "double bind": pursuing economic success, but doing so in an environmentally responsible and legal way (ibid.: 106). On top of this comes the layered irony that, considering the global environmental impacts of different forms of tourism, ecotourism has been found to have a larger "ecological footprint" than mass tourism (Marzouki, Froger, and Ballet, 2012). These binds are indicative of a conflict between popular conceptions of environmental protectionism and conservation and people's contemporary lifeworlds. The former are commonly advanced and reinforced by privileged ecotourists and related governance mechanisms that make up a form of "green citizenship." In addition, and often in direct contrast, are the realities of the local people involved in farming, fishery, and service industries, who are often portrayed as environmental spoilers.

Qualitative attention to the corporeality of contemporary food in the Galapagos island of San Cristóbal leads us to ask: what potentialities do ways of producing, exchanging, and eating food hold for the Galapagos? Our interest in identifying these potentialities is not prescriptive (i.e., we were not immediately preoccupied with finding solutions), but rather exploratory: we want to better understand the creativities and multiplicities embedded in people's living and being. We therefore utilize people's daily food as an analytic device to better understand the potentialities in the constitution of existing practices for addressing democratic tensions between conservation interests and livelihoods in the Galapagos, in particular on the island of San Cristóbal.

From green to corporeal citizenship

At the core of democratic tensions lies the concept of citizenship – understood as a bundle of relationally defined rights and responsibilities. In the context of the Galapagos, constructions of these rights and responsibilities are defined in relation to the previously described double bind. When it comes to pursuing economic activities in an environmentally responsible way, the notion of green citizenship is evoked. In his "conceptual architecture" of citizenship, Dobson (2003) emphasizes how the privilege to "think green" lies central to the ability to achieve the "good life." In contrast, the poor and vulnerable are not as well positioned and are not held responsible for environmental protection. Dobson describes a liberalizing, power-concentrating, asymmetric globalizing world in which agency is exclusionary, in that certain privileged publics, such as the formally educated and wealthy publics that can tour the Galapagos, are assigned greater responsibility than others, namely the politically marginalized publics, such as the colonists that permanently occupy the modern paradise. People who can afford "green" activities (i.e., ecotours of the Galapagos) are understood as having not only the ability to imagine a green lifestyle, but they are also held to own a status of accountability (or being able to claim accountability) for an ecological footprint, while those who cannot afford such activities are described as free of such responsibility. In the process, the poor are marginalized and excluded from the political process. Further, the impacts of their survival-based resource exploitation are understood as different from the environmental consequences of the activity of the wealthy and affluent. In assigning green citizenship to those who can afford it, a portion of the citizenry becomes disembodied from an aspect of public life.

By questioning the epistemological stance in citizen-based environmentalism, we share with Gabrielson and Parady (2010) a call for an ontological shift towards "corporeal citizenship" – taking into greater account the body and the material world as distributed across materialities, where agencies emerge through practices and assemblages that act on daily intentions as well as contingency. Focusing on corporeality reminds us that bodies give substance to citizenship and citizenship matters for bodies (Beasley and Bacchi, 2000: 337). Following Gabrielson and Parady (2010: 376), we envision bodies as "porous, but resistant, plural and connected," thereby disrupting the Aristotelian binary between nature and culture. This human–nonhuman conceptualization of the body and citizenship provides more realistic insight into messy, contemporary socio-ecological dilemmas, in our case, among food, conservation interests, and tourism in San Cristóbal and the Galapagos.

Importantly, similar to Gabrielson and Parady, we apply Pickering's (1995) view of agency as going beyond the subject to include a co-constitutive feature of an endlessly interconnected and dynamic world. We focus on the practices of people who do the work of growing, selling, and eating food in San Cristóbal, an island of about 7,500 permanent inhabitants. From this positioning, the farmers are embodied, and we remain attentive to not only the particularities of human difference, but also to how their living and being are inescapably embedded in their food practices in ways that provide both form and motion to subjectivity and collective agency. This *ontological* perspective on food broadens the scope of possibilities of the environmental and the social, recognizing that the human–nonhuman is simultaneously social and natural.

People's food in San Cristóbal

In continuation, we explore the corporeal citizenship of farmers in San Cristóbal through their food practices as a means of revealing the materiality of human–nonhuman worlds and their co-constitutions. We present a sample of ethnographic research conducted at three food-related sites in San Cristóbal: (1) agricultural production by farm families; (2) marketing of food at a popular open-air food market (the *feria*), and (3) people's food flavors and tastes. Based on fieldwork undertaken by Franke, we present these accounts as first-person narrative.

Cultivating paradise: the natural history of a farming family

Catalina and Fernando are a young couple, she in her twenties and he almost in his forties. They live and farm on 25 hectares of land in El Árbol Antiguo, located at about 600 meters above sea level in San Cristóbal, where the altitude attracts clouds, moisture and rainfall, thereby leading to soil development and vegetation. Elena, a 60-year-old woman, and her 40-year-old son Santiago, also live in El Árbol Antiguo. They bought five hectares of farmland on arrival in the 1970s, when many people migrated from the southern Ecuadorean Province of Loja due to a severe drought. They came to San Cristóbal in search of "the good life." El Árbol Antiguo is a place where families largely grow food for their own consumption, but some also sell leftover produce in Puerto Baquerizo Moreno, located about 30 minutes away by automobile. Puerto Baquerizo Moreno is the island's capital and home to about 6,000 residents.

Slightly further away in the village of Madera live Angélica and Juan, an elderly couple in their sixties who settled on about 100 hectares of land. In contrast to El Árbol Antiguo, Madera had better access to water, which facilitates planting but also the proliferation of insect pests, such as *carmelitos* – a fruit fly from the mainland that arrived in San Cristóbal in banana shipments. Elena and Angélica's families had migrated to San Cristóbal before they were born, in a time when early settlers in San Cristóbal were able to occupy as much land as they wanted. Having arrived when nearly no ships or planes from the Ecuadorean mainland accessed the islands, they learned how to live from what they could grow or find on the island. While they continued to work in agriculture, their children pursued other forms of income-generating activities in the more populated settlements on the island.

When I am working in the field with Catalina, she says that it would be much less work to focus on meat production and have a little vegetable garden for household consumption. She tells me, "If you have a lot of products of one type, you have to find intermediaries but you need to know how to negotiate." What Catalina refers to is the margin one pays to the intermediary to sell the product to the final consumer. She continues explaining how difficult her first year in agriculture was. "I thought I knew the basics of planting and harvesting and that this would be everything." In the beginning, when she started selling her products at the market, she felt ashamed and found it difficult to smile, greet people, and invite them to buy something from her. "When Angélica does it, it [selling produce] seems so easy, but for me it was difficult to learn." Nowadays, Catalina assures me, she feels much more relaxed about it.

Santiago produces according to the principle: "Of everything a little, a little of everything." He is careful when using synthetic agriculture inputs. He prepares a mixture of poultry manure that he gets for free, when he pays the transport costs of US$ 20. Santiago explains, "The poultry manure alone is not strong enough and the harvest gets bigger when it is used together with synthetic fertilizer." He brings along a bucket half filled with the liquid mixture of synthetic fertilizer and shows me how to give it to the still growing cucumber plants. He tells me, "I use [pesticides] only when 50 percent of the plants are already ill. Not that much and only when the pests are already there."

Angélica states, "People use synthetic inputs because of lack of labor. If there are pests, they take out the sprayer, and the problem is solved." She adds, "We produce 100 percent organic." They tell me they do not apply any synthetic fertilizer, pesticides, or herbicides because it is bad for their health. Juan explains, "I fear the products that come from the continent because they contain chemical compounds. I learned [organic farming] from my parents, because in the past there were no synthetic inputs."

Fernando explains, "Red beets, carrots, alfalfa, and coriander grow without additional inputs. Cassava and papaya are all natural." We go to the part of his field where he has cabbage. He opens the leaves to reveal an early stage snail feeding at the center. "When they eat everything, the cabbage loses its value." He adds that they spray insecticides when the plants are still young to prevent the pests from taking over. I ask him whether insecticides are allowed on the Galapagos, and Fernando shows me a blue-labeled pesticide container. "This is less toxic," he states. "The red label [highly toxic product] is used everywhere on the Ecuadorean mainland, but I prefer the blue one, which is less strong and not as bad for your health."

Juan tells me about his life and farming: "My parents knew that one has to store the seed. We always saved a lot of them, up to ten quintals for the next year." I see Angélica collecting the seeds from the tomato she is preparing for a salad. When I ask her about the differences between her seeds and the ones that one can buy, she informs me, "They are the same, both are good." In contrast, Fernando prefers hybrid seeds and explains, "My parents bought their seeds from a store in Puerto Baquerizo Moreno, but only 40 to 50 percent of the seeds germinate. The germination rate of hybrid seeds is 99.9 percent, but 100 tomato seedlings cost US$ 100." Sometimes he and Catalina order from an agronomist on the mainland. Santiago also has started to experiment with hybrid seeds. One day while in the field we filled plastic cups with soil and we dropped one tomato seed into each cup. In reference to his heavy investments as a farmer, Santiago confides, "Sometimes you lose more than you win."

Marketing food: a day at the feria

It is still dark outside on Saturday morning at 4:30 a.m., when Elena and I arrive at an open-air food market, the *feria*, in Puerto Baquerizo Moreno. The feria is an initiative of the Ecuadorean Ministry of Agriculture (MAGAP, Spanish language acronym). The first open-air food market was held in April 2015, with the objective of better balancing imported products with what farmers in San Cristóbal can grow.

The taxi pickup truck that picked us up on the farm stops at the entrance of the street where the feria is located. Upon arriving, I start to unload the boxes packed with produce for sale. The previous day, we harvested oranges and tangerines, plantain, papayas, tomato, lemons, cassava and coriander. A man approaches us to load boxes on his pushcart to bring them to Elena's makeshift stand. No customers have arrived yet, but I spot a market saleswoman approaching from across the market. In general, the market sales people are almost always female, and are not farmers, but intermediaries who buy and sell for a profit.

The saleswoman arrives to ask, "Sell me your tomatoes for US$ 5?" "No!" Elena replies. "Only for US$ 3, not [a penny] more." In a moment, Catalina arrives with three wooden boxes of tomatoes. The saleswoman takes the amount of tomatoes offered to her and leaves. A second saleswoman arrives at the stand and asks, "Papaya?" "I have a whole box for twelve dollars," Elena says. "I do not want that much," the saleswoman replies. She takes three large papayas instead and carefully selects a few other minor products.

Opposite the stand is a more elaborate stall owned by a saleswomen. It is a wooden construction, painted blue. Shelves line the walls filled with onions, powdered milk, coffee, cooking oil, cans of tuna, and plastic bottles of shampoo. There are two refrigerators with soda, cheese, and sealed packages of chopped vegetables. Further to the front is a blue scale, and to its right lay packages of rice and ground maize. Boxes filled with fresh, unprocessed vegetables are displayed up front. Angélica and Juan walk through the street to reach their table next to us. Angélica carries ten cheeses and they have papayas as well.

Santiago explains, "In the past, everyone had the same. There was nearly no market because everyone grew food … We had five donkeys to carry our harvest. There were practically no roads and there was a lot of rain … Now, there is a market and there is even competition between the products." Elena adds that they bought what they needed, such as oil and salt, from the money they earned. "Now there is business because of tourism. One can sell a lot and in the same way one can produce a lot. Cucumber, for instance. We almost did not cultivate it because there was no one buying it."

I talk to an older woman at Elena's stand, who ordered a chicken from Elena's farm the day before. "I buy a bit from everyone," she explains and adds, "In season, products are fresh, such as tomatoes. I also buy [mainland produce] from the stores, because the plantain from here is much more expensive. The price of the farmers is not fair because they are from here and products coming from outside need to incorporate the transport costs." I go to Catalina and Fernando's table, which is at the other end of the row of tables and blue-and-yellow tents arranged by the staff of MAGAP. I see a man with a baby in his arms and who is buying chard and other vegetables. When I ask him about the quality of the produce, he answers, "They are all the same. Quality and price are all the same." He continues buying his groceries, when I approach another woman. She says, "It is cheaper to buy from local farmers, but the quality is not always good. The products from the mainland are better. I buy bananas, papayas and oranges from here because they are fresh, but I prefer to buy them from the stores."

When I stayed with Catalina, she told me, "Our product does not have value because it is produced here." She added, "They say you do not do anything, that my actions of weeding and maintenance do not cost anything and that they therefore do not understand why our products are not sold for much cheaper." This reflected a perception of clients who believed that the imported products need to take into account the shipping costs. Fernando made this comparison as well when he stated, "Imported products are very expensive and often are twice the price because of arriving by ship or plane."

Santiago observed, however, that there were also people showing more interest in products that came from San Cristóbal. He provides an example, "A salesman told me once that he could not buy from me because of my high price; he could get the box of cassava for US$ 6 from the mainland, whereas has to pay me double the price. A week later, the salesman lost money because the cassava from the continent had arrived rotten." Similarly, a saleswoman at the market said that she preferred to buy from local farmers, but they did not have enough supply." The saleswoman added, "When I order [produce from the continent] at 8.00 a.m. in the morning, it arrives here at 2:00 p.m. on the same day, and it is of good quality, large and very cheap." I asked her why she thought people did not supply enough, to which she responded, "People do fishing and other types of work. Even though the soil here is very fertile, they do not cultivate."

Experiencing food: tastes and preferences

It is 6:00 a.m. in the morning, and I am helping Catalina prepare rice and coffee for breakfast. "It is later than usual" she says. Her son did not have to go to school, so she is still in her pajamas. There is an open space in the house that includes a common living room and the kitchen area to the right. There are packages of cereal and Italian pasta next to other food products stacked under the kitchen counter. The father left some fruit on the counter that he had collected from the field. I start cutting a papaya, and ask Catalina whether she also would like some. I am astonished by the diversity of fruit that they grow. When I confide that I usually just have fruit and yoghurt in the morning, Catalina responds, "I should have a light breakfast more often. I want to lose weight."

Meanwhile, Angélica declares, "I know what tourists like to eat," which leads her to prepare a small salad for just about every lunch. "Usually Juan does not eat salad at all, but I told him to set a good example," she says. Juan asks me whether I had tried the food in Puerto Baquerizo Moreno. He explains, "It is bad food because it has either too much sugar or it is from bad [quality] meat." When preparing ceviche – a popular raw seafood dish cooked in pure lemon juice, they mix the kitchen water and put lemon to it." Juan adds jokingly, "I taught Angélica how to prepare ceviche [the right way]. I learned how to do this during my time as a fisherman."

When I ask what the differences are between the products that come from the Ecuadorean mainland and those grown on San Cristóbal, Fernando responds, "They seem to be fresh because they are still frozen from the transport. What we produce is fresh because it is [harvested] the same day, whereas the other products are harvested around eight to 15 days before and are in poor condition." Fernando tells me this by referring to his experiences as a trader, where he saw how chemicals were applied to food products to regulate the ripening process and assure storage. Likewise, Juan argues that there is an increased amount of chemicals applied to the food imported from the continent. He deplores the foods that come off the ships, arguing, "They [the foods from the mainland] cause diseases, such as cancer … But the people from the urban area prefer those products, because they deteriorate less quickly than the locally grown ones." Angélica expressed a similar sentiment, "They [the fruits and vegetables] apparently seem beautiful, but you cannot really see what you are eating, and you do not know what you are eating." To prove her point, she showed me a potato that she had purchased from the mainland, pointing at its rotten core.

Catalina leaves the field at 11:00 a.m. to prepare lunch, which allows her to avoid being outside during the hottest part of the day. Before we go down to their house, she cuts a lettuce, which we take with us. "From plant to pot," Catalina says proudly. In the meantime, Fernando is about to harvest some tomatoes for the feria, when Catalina stops him, "If you take the few that are ripe, I am left without any tomatoes for the house." She explains, "The tomato is indispensable at home. For every dish you will need tomatoes." I ask about the difference in taste, and she replies, "It is healthier and the taste is different because it ripens on the plant. It has a pleasant taste. It can be mild, not rubbery or watery. Tomatoes from the mainland can be very hard because they have been ripened by force, or they can be very soft if they are too ripe already." In contrast, tangerines are not imported from the Ecuadorean mainland. Holding a tangerine in his hand, Santiago tells me, "There are many varieties here. There is one that is quite big, and it tastes like an orange, but the skin is harder." He continues, "The whole year around there is production of tangerines and oranges. They even transport them from here to other islands of the Galapagos."

Elena always shops for groceries at the same stall on the market after the feria has ended on Saturdays. The saleswoman of the stall asks her to sit down, and Elena gives her a list of her needs for the week: toilet paper, cleaning material, onions, garlic, pasta, sugar, corn flour, Maggi seasoning, napkins, two packages of milk and two bottles of cooking oil. Elena also buys 15 pounds of *mote* (pre-boiled maize), which she will cook and put into packages for resale at the feria. Elena explains, "This type of mote does not exist on the islands. It can be produced here, but it stays small and the people do not buy it." She adds, "This is why I do not produce it anymore. Our mote is sweet and fresh, but the people want to have big mote." Elena processes the mote from the mainland because she sees a lot of demand for already cooked or chopped products. Elena explains that even the salespeople buy it from her.

While Angélica and Juan shop for groceries at sundry supermarkets in Puerto Baquerizo Moreno, Catalina prefers one that is two streets higher up from where the market is located. "It is cheaper, and if I miss something during the week, I can give them a call and they arrange to transport it to El Árbol Antiguo." Catalina buys the things that are not available on her farm. She assures me, "It would be absurd to buy onions, if we have them here." At the same time, Catalina explains that she prefers the chicken that is imported from the mainland: "I like the taste better of the chicken here, but the imported chicken fed with chicken feed mixture cooks much faster. Instead of an hour, it takes me 30 minutes to prepare chicken for breakfast."

Discussion: feeding paradise

While the Galapagos hold a great deal of international intrigue and wonder as a place to be preserved, in practice, denizens such as Catalina, Fernando, Angélica, Juan, Elena and Santiago as well as salespeople, food shoppers and others who establish a home and make a living in San Cristóbal bring forth a very different imaginary of "paradise" through their co-constitution of food and daily life, with the surrounding landscape and wildlife. For instance, when Catalina describes the place where she lives, she says, "In San Cristóbal you can find everything." She likes the tourist sites that are "beautiful," but she also complained to a taxi driver one morning about the conservationist strategy to eradicate goats. Goats are shot because they represent one of the most destructive invasive species on the Galapagos, eating the natural flora while adapting easily to the conditions on the islands. Reflecting on the killing of the goats, Catalina said that it "makes life here even harder [for farmers], because they [goats] provide additional income and actually eat the weeds. It is difficult to cultivate the land if it is left full of blackberry [plants]."

We initially presented a "bind" created by two competing visions of the Galapagos: a site of untouched nature and a site of livelihoods. Epistemological privilege and ways of thinking about citizenry have perpetuated this divide through the marginalization of an important number of local actors. Given that the environment is involved in shaping the subjective experience of food, we argue that Dobson's "green citizenship," which neglects nature–human interactions, is fundamentally exclusive to an elitist notion of environmentalism. A focus on corporeal citizenship opens opportunities to revalue matter and things, which broadens the scope of ways of being sustainable. For each of the farm families described in the vignettes, the food they grow is part of their daily embodied experience of eating and, through it, being able to know what one eats and exchanges at the market. Elena explained that her family has started using fewer chemicals for their personal health, but also to avoid harming other people on the island. The other two families do so as well, each employing a diversity of strategies for using fertilizers and pesticides and obtaining seeds within ranges of what they believe is good for them and for San Cristóbal.

The farmers explained that they farmed to save money on food that would otherwise have to come from the mainland as well as to earn some income, although jobs in fishing and tourism paid better. They experience the double bind of economic success and environmental citizenship in their daily lives, often in confrontation with others. Fernando explained, "They [the authorities] see the Galapagos as part of tourism. They want to get rid of us, so that we do not pollute, plant or raise animals." This tension was also observed at the feria, when consumers expected to pay lower prices for food produced on the island. At the same time, when talking about their experiences, these farmers reveal an appreciation and personal passion for being able to grow, taste, harvest, and sell fresh food – food that is linked to the island and to their labor. The work they do – with their bodies, on the land that they farm, in the kitchens where they cook and eat, and in markets where they sell their food – reaffirms that bodies give substance to citizenship and citizenship matters for bodies (Beasley and Bacchi, 2000: 337). It is through these embodied practices of food that at least three potentialities emerge.

For 20 years, biologists from the Charles Darwin Station and, more recently, policy makers at MAGAP, have identified local, responsible farming as an opportunity for environmental protection in the Galapagos. For tourists and the tourism industry, local food production is widely perceived to be a threat to a pristine Galapagos – an "eyesore" on Darwin's legacy. Nevertheless, imported food has been identified as a primary source of invasive species, in particular insect pests and diseases, to the archipelago (Galapagos National Park, 2009a). In addition to disturbing the natural ecology, invasive pests are a threat to the viability of agriculture on the islands. In an effort to avoid imports of invasive species, authorities at the Galapagos National Park and MAGAP have decided to promote island-based food (Galapagos National Park, 2009b). Thus, despite the historically antagonistic socio-environmental context of the Galapagos, a *first potentiality* of decreasing invasive species emerges from the innovative agriculture of rural families in places such as San Cristóbal.

A dependence on imported food from the mainland is a constructed reality and embodiment that goes against the explicit environmental interests of tourists and that is continually being undermined by island-based farmers and fishers. As recently proposed by MAGAP, substantial artisanal fishing and small-scale farming in places such as San Cristóbal could come to feed the permanent residents as well as the tourist industry, thereby displacing the tourist industry's near total dependence on food imports. In the process, a *second potentiality* of decreasing reliance on fossil fuels tied to shipping can also be achieved by local food production, while opening the possibility for a third island-based potentiality.

The open-air food market (or feria) provides a space where farm produce is sold under the slogan: "I commit myself to being a farmer who thinks about the future of the people and the islands by learning about more ecologically friendly production techniques and offering a healthy, fresh and natural product." While new foods are being added to the variety of choices and preferences of farm families, such as the imported chicken that cooks faster and tourist-friendly dishes (e.g., the green salads of Angélica), embodied food practices are deeply related to locality. The embodiments of imported food with regard to form, shape, flavor, and availability on demand also create a counter tendency in the form of an alternative market for island-based food that is fully appreciated for its freshness and flavor as well as for the possibility to directly link consumer interests with source and means of production. Freshly caught fish and organically grown citrus fruits, tomatoes, and plantains are just a few options that are essential for the daily life of people living in and visiting the Galapagos. Local food consumption, cuisine and gastronomy, and knowing how products are grown, harvested and cooked is a *third potentiality* that could be fostered by emphasizing a certain local "taste" in healthier, more responsible Galapaganian food, thereby further reducing the ecological footprint of tourism.

Conclusion: corporeal food citizenship

We opened this chapter by describing a bind of economic success and environmental citizenship to summarize an ongoing tension in the Galapagos. The vignettes present diverse human–nonhuman constitutions of food: sites and practices that effectively broaden the possibilities for environmentalism in the Galapagos.

Despite what may be perceived as faults in their "green thinking" by virtue of their limitations, farmers in the Galapagos tend to be intrinsically, rather than instrumentally green, in the sense that their food production, circulation, and consumption tend to be resource limited and conserving. Based on our empirical study of contemporary food practices in the Galapagos, we identified three potentialities that rest in certain island-based food practices vis-à-vis the competing embodiments of tourists, business people, residents, plants, pests, scientists, and park officials. Shining light on the neglected promise of people's food in San Cristóbal highlights the value of studying corporeal citizenship to deepen transdisciplinary thinking and multifunctional politics and action around the possibilities of responsible living.

The latent potentialities present in people's daily food practices in San Cristóbal lead us to suggest scholarship capable of accommodating the diverse, seemingly contradictory concerns that not only relate to human health, equity, and the environment, but that also cut across the intersectionalities of people's daily living and being. Greater attention to the richly heterogeneous corporeal realities of food can help identify existing, proven prospects for overcoming the dualities that divide conservationists and colonists in the Galapagos, while opening space for new, more inclusive green citizenship.

References

Beasley, C. and Bacchi, C. (2000). Citizen bodies: embodying citizens – a feminist analysis. *International Feminist Journal of Politics*, 2(3), 337–358.

Brewington, L. (2013). The double bind of tourism in Galapagos Society. In S.J. Walsh and C.F. Mena, *Science and Conservation in the Galapagos Islands: Frameworks and Perspectives*. New York: Springer, 105–125.

Dobson, A. (2003). *Citizenship and the Environment*. Oxford: Oxford University Press.

Gabrielson, T. and Parady, K. (2010). Corporeal citizenship: rethinking green citizenship through the body. *Environmental Politics*, 19(3), 374–391.

Galapagos National Park. (2009a). *Invasive Species in Galapagos*. Available at: www.galapagospark.org/nophprg.php?page=desarrollo_sustentable_especies_invasoras (accessed 2 June 2016).

Galapagos National Park. (2009b). *An Area in Need of Renovation*. Available at: www.galapagospark.org/nophprg.php?page=desarrollo_sustentable_agropecuario (accessed 2 June 2016).

Marzouki, M., Froger, G., and Ballet, J. (2012). Ecotourism versus mass tourism. A comparison of environmental impacts based on ecological footprint analysis. *Sustainability*, 4, 123–140.

Pickering, A. (1995). *The Mangle of Practice: Time, Agency, and Science*. Chicago, IL: University of Chicago Press.

Quiroga, D. (2009). Crafting nature: the Galapagos and the making and unmaking of a "natural laboratory." *Journal of Political Ecology*, 16(1), 123–140.

Taylor, J.E., Hardner, J., and Stewart, M. (2006). Ecotourism and economic growth in the Galapagos: an island economy-wide analysis. Working Paper No. 06-001. Davis, CA: Department of Agricultural and Resource Economics, University of California, Davis.

Wolford, W., Lu, F., and Valdivia, G. (2013). Environmental crisis and the production of alternatives: conservation practice(s) in the Galapagos Islands. In S.J. Walsh and C.F. Mena, *Social and Ecological Interactions in the Galapagos Islands*. New York: Springer, 87–104.

9

UNFOLDING AGENCIES AND ASSOCIATIONS OF AGROECOLOGY NETWORKS IN BRAZIL

Flávia Charão-Marques, Claudia Job Schmitt, and Daniela Oliveira

Introduction

In recent years, concerns related to sustainability in agriculture and food production in Brazil were incorporated into the public agenda, including several prominent policies addressing hunger and promoting family farming.[1] It would be a mistake, however, to assume that this evolving agenda is exclusively a result of state technocratic intervention. In the case of agroecology movements in Brazil, the social engagement goes back to the 1980s, when the initial contestations in favor of an "alternative" post-industrial agriculture encompassed the denunciation of the well-documented harmful effects of pesticides and damaging social impacts of modernization (i.e., concentration of land tenure and deepening inequalities). This contentious debate emerged as an expression of a diverse community of farmers, "foodies," and social activists and became increasingly visible in public arenas. From the year 2000 onward, agroecology movements strengthened ties with civil society organizations, advocating for "food security and nutrition" and the "human right to adequate food," and in the defense of public health. Most recently, discourse has diversified to include "responsible consumption," "food quality" and "prevention of obesity and noncommunicable diseases."

In terms of political activism, the first national efforts to connect local initiatives started with the Alternative Technologies Project (PTA)[2] in 1983, and after with the National Agroecology Alliance (ANA), usually identified as a network of networks,[3] that came together in 2002. Nevertheless, the core of agroecology networks does not reside solely in the political realm. It is important to highlight, first, spatially bounded relations (and site-specific ecological processes) as crucial elements of this activism. Second, these networks result from continuous translation[4] and experimentation of a plethora of practices, understood as open-ended sets of doings and sayings, inherently spatial, largely bodily performed and interacting with material arrangements (Schatzki, 2015). In the same direction, Law and Mol (2011: 274) argue, "Perhaps associations are not just a matter for social beings, but also one to do with materials." Moreover, we are considering that the change in agriculture and food is not the outcome of highly unidirectional and predictable development processes. They are rather part of multiple (Woods, 2007), mutational (Arce and Long, 2000) and contingent (Paredes, Sherwood, and Arce, 2016) relationships and social–material processes, which potentially create embodied subjectivities (Blackman *et al.*, 2008).

From this perspective, the expansion of networks can be described as a wide spectrum of contingent associations and interactions, creating multiplicities, which probably affect flows of change. Based on long-term research and personal involvement, this chapter explores how critical and practical action against dominant modes of agriculture and food consumption fostered creative associations between practices, agencies, and materialities to give birth to agroecological networks in Brazil. We ask: How do emergent agencies and potential associations make room for agroecology networks? Which processes contribute to enlarging the networks under particular contingencies? Do situated practices contribute to political activism?

Approaching the empirical

We would like to talk about land, seeds, food, practices, and social change. For this purpose, we explore the trajectory of two ongoing experiences in Brazil, situated in *Serra Gaúcha* and *Zona da Mata*, where agencies, creativities, and different associations emerge from practices connected to food production. The basis of our argument in this chapter comes from different empirical studies conducted between 2009 and 2015, involving background document research, participant observation, and ethnographic interviews. Authors' direct experience with the agroecology movement and the networks dates from the 1990s. Using a multi-sited ethnographic approach, we explored a variety of connections linking scattered processes in space and time to follow the two situated experiences.

Research was conducted in the Serra region, in the state of Rio Grande do Sul, located in the southernmost part of Brazil, as well as in Zona da Mata, in the southeastern region of the state of Minas Gerais. While 2,000 km separate these two interconnected experiences, they were selected because of their importance to the Brazilian agroecology movement, as they are among the earliest and most influential initiatives emerging in the country.

The two cases express different phases of political activism and situated practices, assembling objects and subjects in different flows of change. In the Serra Gaúcha network, the earliest problematization (Callon, 1986) was the effect of pesticides, which evolved into a critical stance against genetically modified organisms. Taking place in different markets, relations established with consumers enabled the emergence of participatory certification as an alternative to conventional certification schemes. In Zona da Mata, agroecology became part of a broader movement for farmers' rights, including sharecroppers, women, and other people threatened with being displaced by nature conservation units.

The experiences have commonalities and differences, but it is important to clarify that the main goal of our research was not to identify a definite set of structural differences or similarities between the two cases. As suggested by Law (1994), we tend to engage in a sociology of verbs rather than a sociology of nouns, understanding the social world as "this remarkable emergent phenomenon that, in its processes, shapes its own flows" (ibid.: 15). In doing so, we did not seek to map the organizations involved in the respective networks. Instead, we applied our qualitative lens to the constitution of network relationships and assemblages (i.e., the processes, effects, and affects of networking), in the hope of avoiding representations, first, of networks themselves, second, of agroecology, and third, of food. As such, the study seeks to reveal the intentional, unintentional, and even accidental relationships that people establish when they are experiencing multiple changes in food production as well as in their own social world.

The emergence of ecological agriculture in Serra Gaúcha

> [T]he objective was to experiment with another model of agriculture ... to test things from another model, without chemical inputs. This idea emerged in the Legislative Assembly during discussions about the agrochemicals law. People used to say, "Ah, but there's no proof; you don't know if this will work; there are no concrete examples; there's no one who has done this. You can play around in your little orchards without poison, but you cannot do serious work." This was the argument of those who were against the regulation of agrochemicals, and we were sued by the industry and its representatives during the discussion of the law. We were required to present concrete examples. This was the idea behind the Vacaria Project.

This is how agronomist Maria José Guazzelli describes the Vacaria Project, the initial location of ecological agriculture in the Serra region. In about three decades, a diversified network comprised of farmers, small local agro-industries,[5] *ecological food*,[6] creole seeds, booklets, manuals, and different technical publications would be operating in different municipalities, with connections at the national and international levels.[7]

In Serra Gaúcha, Italian immigration from the nineteenth century onwards had a strong cultural influence, contributing to the establishment of colonial forms of agriculture with diversified food production. In this region, family farmers established relations with different markets even before agricultural modernization, trading lard, wheat and wine with different regions of the country. Some of these farmers were able to accumulate capital and invest in the industrial sector (e.g., metal-mechanical and furniture industries). Over time, Serra Gaúcha became the most traditional grape-producing region in Brazil, producing wine and sparkling wine on an industrial scale, and also developing a tourist complex associated with these activities. Land ownership were retained by family farmers, and the region was considered "developed." Nevertheless, social and economic inequalities were enhanced with the dissemination of modern agricultural techniques after the 1960s. These negative effects were compensated somewhat by the migration of workers to the flourishing industrial sector and by pluriactivity.

The Vacaria Project[8] began in the 1980s, emerging from environmentalists' engagement with the struggle against pesticides that started in the previous decade.[9] They created a center that aimed to "make concrete" a different model of agricultural production initially focused on the demonstration of and experimentation with alternative farming practices, seeking to oppose the notion that it would not be possible to produce food without agrochemicals on a commercial scale. At that moment, there were no experimental programs for sustainable agriculture in Brazil. Traditional agriculture was regarded as backward and the productivist model was considered the best response to challenges in the region.

The technicians involved in the Vacaria Project drew on European experiences in biological agriculture, particularly the work of Francis Chaboussou,[10] with the support of international networks and development agencies willing to fund sustainable initiatives. These agencies were interested mostly in learning from grassroots experiences, allowing significant space for experimentation and discovery. The central axis of the work developed in Serra Gaúcha was *to produce food without poison.*

From 1985 on, the agronomists started to interact with local communities through the Rural Youth Pastoralist (PJR),[11] facilitated by a priest, Father Schio. Young farmers were concerned about rural exodus and recurrent episodes of intoxication by pesticides, especially among apple producers (the first commercial crop modernized in the region, followed by grapes and vegetables). In some cases, farmers affected by health problems were already starting to learn about natural food and macrobiotic practices, but this was not a general feature of the group.

Young farmers from PJR were influenced by Liberation Theology and held a critical position in relation to modern agriculture, but it is clear that there were practical motivations for establishing closer ties with the project. Ecological farming represented an ideological resistance, but also the possibility to escape from the exclusionary tendency of modern agriculture, particularly due to asymmetrical relations with different market agents. The connections established between the PJR and the Vacaria Project created an enabling environment for farmers to envision new possibilities:

> What did I have? I had a very strong conviction … it was clear, we didn't want this for our family, for our children, for our society. We didn't want this type of model in which people get sick, or have the soil and water contaminated, and all of that was inside us. But we had an ideology, and we had a vision, we had principles, the principles of life, no one can beat the principles of life. And that's why we started seeking alternatives.
>
> *(Male ecologist farmer, one of the leaders of PJR)*

Changes in practices were associated with the reconfiguration of daily activities and social relations inside and outside the family:

> At that time we felt insecure, so we, including myself, made a little compost, had an experience here and there. I started working with peaches. After that, I started to phase out the use of pesticides on my apples. I was scared to dive right in. It seemed that it could work but I was afraid to fail. My father was following me and my ideas; when something went wrong, he would be very upset. I affected the whole family …, so everyone started doing small-scale experiments, in different places on the farm. When they became more confident, they started to produce ecologically.
>
> *(Male ecologist farmer, living in the Serra Gaúcha region)*

Relations established with *Coolmeia*, a consumer cooperative located in Porto Alegre, the capital of the state of Rio Grande do Sul, enabled the organization of the first ecological farmers' market. Farmers traveled about 200 km once a week to sell their products in the city. They started to establish new ties with environmentalists, food activists, and responsible consumers. Gradually, different farmers' markets were organized in Rio Grande do Sul supported by several counties. The farmers also traded products in small shops and supermarkets in different parts of the state and the entire country.

New connections emerging outside the farm stimulated farmers to develop new products and they started to process grapes, tomatoes, and different types of fruits using artisanal techniques. Later, they evolved to the construction of small-scale multiprocessing units, belonging to one family or to a small group of farmers. Step by step, situated practices emerged, as stated by a farmer who has a small agro-industry, looking back on this process:

> The techniques, processing and the market encouraged people to engage in ecological agriculture. These techniques were possible because half a dozen farmers started to apply them on their farms. Through on-site practices where agronomists and producers were working together, these techniques evolved.

Ecologist farmers and *ecologically based agriculture* emerged in the Serra Gaúcha network as locally constructed concepts.[12] The terms were negotiated and chosen by technicians, farmers, and consumers to make explicit their opposition to organic farming, which they understood as a narrow political strategy oriented towards input substitution and the creation of niche markets. This construction generated different forms of political activism. In the 1990s, network organizations were active in mobilizing against the commercial use of genetically modified organisms in Brazil. In the beginning of the next decade, *Centro Ecológico* and farmers' organizations from Serra Gaúcha played an important role in the public critique of conventional forms of certification of organic products during the formulation of the Brazilian Law for Organic Products. One outcome was the emergence, in southern Brazil, of *Rede Ecovida de Agroecologia*[13] which, in partnership with other organizations, achieved state recognition of its participatory certification, and the creation of an innovative legal framework in the country that provided more space for multiple expressions of sustainable food production.

Agroecology in Zona da Mata

Critical action against modern agriculture emerged in Zona da Mata in the 1980s, resulting from the confluence of the political struggles of the Rural Workers Unions (STRs) and the alternative agriculture movement.[14] Links established between these two streams of contention and *the peasants' life worlds* created an enabling environment for new creative associations, weaving struggles for the recognition of farmers' rights (including access to land) and the transformation of food production together. This took place through the development of agroforestry, agro-homeopathy, home gardens, as well as the expansion of farming oriented schools, changes in gender relations, and other practices.

> How can a peasant who does not have access to land work with agroecology? He does not want to use pesticides, he is aware of that, but the landlord says he must spray pesticides. … To practice agroecology people need to have access to land. The struggle for land is part of agroecology. The labor unions started to make visible the suffering of the sharecroppers. They had an ecological conscience but did not have the right to choose what kind of practices they wanted to use.
>
> *(Agroecological farmer, former union leader in Zona da Mata)*

To better understand the link between agroecology and peasant struggles in Zona da Mata, it is necessary to retrace its history. Up to the early nineteenth century, the Portuguese colonial government defined the territory as an interdiction zone to prevent gold smuggling from mines in central Brazil to the coast. Belatedly, agriculture expanded. After the mining decline, the Atlantic Forest was cut down to make space for coffee plantations. Slash-and-burn techniques rapidly depleted the soil. Nevertheless, diversified peasant agriculture survived in the fringes of the coffee economy, producing food for self-consumption and some surplus for trading. After the abolition of slavery at the end of the nineteenth century, European immigrants were attracted to the region, joining poor farmers (descendants of the Portuguese, Africans and native indigenous peoples) as part of the labor force. Coffee producers adopted diverse modalities of sharecropping arrangements (*parceria*) as an alternative to slave labor. Plantation owners permitted the intercropping of coffee with beans and corn for self-consumption, but most farmers had no ownership of land (Lanna, 1985).

Cyclical crises in coffee prices drove the diversification of farming, the emergence of commercial agriculture and the fragmentation of large landholdings. At the same time, after the 1960s, modernization policies in the region addressed mainly coffee production. Pesticide use and large debts came along with the continual subordination of sharecroppers and smallholders to landowners[15] (Cintrão, 1996). This situation led to the creation of unions, supported by the Pastoral Land Commission (CPT), the Good News Movement (MOBON), and Ecclesiastical Grassroots Communities (CEBs).[16] During the same period, a group of agronomy students at the Federal University of Viçosa (UFV), located in Zona da Mata, organized the Viçosa Alternative Agriculture Group (GAAV). Over time, linkages were established between STRs, the GAAV and the PTA.[17] This convergence led to the creation of the Zona da Mata Center for Alternative Technologies (CTA-ZM) in 1987.

Initially, young agronomists working at the CTA-ZM focused on experiments and advisory services related to alternative farming practices, such as the recovery of traditional varieties, ecological soil management, and new techniques of animal husbandry. Between 1980 and 1990, the crisis in the international coffee prices contingently favored the search for diversification of farm activities, reinforcing the role of CTA-ZM and, later, researchers from UFV. Technical staff was also involved in political and organizational activities, helping union leaders present legal claims and championing the rights of rural workers in the Brazilian Constitution, as well as creating new markets for rural farming families. Despite continuous threats from opposing interests, these ongoing, intertwined processes of farming practice, social organization, and political action continue to shape the network today.

The network broadened its scope and actor enrollment, but the rise in coffee prices combined with the strengthening of Brazilian agribusiness from the 1990s onwards applied new pressures on the territory. These pressures were characterized by an intensification of the use of chemical fertilizers and pesticides. In response to the aggressive activity of agri-sales companies in the region, farmer organizations launched the "Movement in Defense of Life and the Environment." The way they would, in their words, both "denounce" the harmful effects of pesticides and "announce" alternatives represented an effort to go beyond criticism and develop a future project. During the same period, the right to land became, once more, a central issue for this agroecology network. The potential expropriation of land for a nature conservation area as part of the Serra do Brigadeiro State Park, questioned the land rights of seven hundred rural families. After a long process of protest and negotiation, which included a commitment to agroforestry and other agroecological practices in a buffer zone, the park was redesigned to assure the land rights of the families. A decade later, one affected family summarized its experience:

> If I look at my house, my family, our way of life changed about 70% with regard to caring for the environment, caring for food. A different way of seeing the world, humans, and treating people. Our community is also changing, slowly, knowing things that they did not know. Today, almost everyone has a home garden [as a means] to avoid buying products in the market. These changes seem very small, but when you put them together, they change a family and a community.

Out-breaking associations, densifying practices

The trajectories of situated experiences presented in the previous sections share certain commonalities, such as the influence of the Catholic Church's Liberation Theology, and the criticism of modern food production based on a wide range of industrial era practices. These multiple recombinations favored the reformulation of courses of action, which can be identified as counter-tendencies (Arce, Sherwood, and Paredes, 2015). They are not random transformations, but changes deliberately aimed at exclusion and subordination, whose consequences are both social and material. Empirically, the counter-tendencies were enacted and materialized in quite radical changes in practices of producing, processing, and trading foodstuffs. Besides, in the formation of coalitions between agents to cope with antagonistic public policy (e.g., in favor of genetic modification technology or "food safety" extremism) or to address specific claims to the state, as in the case of Zona da Mata, to ensure farmer landholding as part of the conservation unit implementation. Coalitions were often created to resist or overcome coercive institutions that organized and protected conventional food production, such as in the Serra Gaúcha region, in order to overcome barriers imposed by health authorities such as radical safety standards in food processing and packaging.

Likewise, farming practices were the first focus of both networks, beginning with the struggle against pesticides, but the process was different in each case. Ecological food identity was more important to the Serra network, which got involved in eco-fairs and later in certification. One pioneer farmer in the fair told us how he experienced these activities:

> Participating in the fair, we started to have contact with other people … who buy at fairs. So, we cease to be that family that used to produce and deliver to the cooperative, … breaking the relationship with our own product. [With the fair] You are the one who sows, produces, harvests and sells, right. So, the fair was a 180-degree turn.

In the Serra we heard, "We entered for money. If it weren't for [the monetary value of] agroecology, we would have gone to the city." But in the Zona da Mata, alternative agriculture has meant a chance to face the historical problem of land tenure. For instance, in a locality called Araponga, a group of farmers, with union support, established a collective fund to jointly purchase land.[18]

Actors in the two networks mobilized their agency in relation to the strategic manipulation of a network of social relations (Long and Van der Ploeg, 1994), channeling specific elements (i.e., kinship, economic contingencies, technology, or policies) and making them interact. The agency is revealed in increased influence over "the others" to modify established states, due mainly to their organizational capacity, in this case, within an often antagonistic political environment favoring industrial food. These forms of ordering can be seen as a product of social agency performance. Nonetheless, producing food is inherently a material process; the result of a transformation of light, soil nutrients and water. This challenges us to consider the socio-material associations.

The agroecological networks support sustainable innovation for local expressions of agroecology, for instance by facilitating the flow of critical factors of production. Objectively, the networks link practices including farming, marketing, food processing, organizing, advising, experimenting. We cannot reduce the growing interactions among these activities to a normative arrangement. Rather, it involves the densification of creative arrangements and subjectivities through the overlapping of different actors and unpredictable emergent associations. Below, two brief vignettes illustrate the potential of such "out-breaking" associations.

Box 9.1 A seed farmer at Capela de Santo Antão, Serra Gaúcha

Exiting the highway and arriving at the countryside or *colony*[19] in Brazil feels like time travel. Nineteenth century wood houses, vineyards, flowering roads and people who speak with an Italian accent are remarkable elements, and often even a tourist attraction. This bucolic scene is also part of the daily life of the Menegat family, who talk about work and its intimacy with nature, since their ancestors used to grow wheat and corn, and raise pigs. Vilmar Menegat summarizes the past: "The land was worn out." The only way to make a living was to cultivate apple trees for recently established packing houses, but the family was afraid of "banks and poison." A fair located in the state capital was organized by the *Centro Ecológico* and *Coolmeia* and proved to be an important stimulus for ecologically based farming and the organization of rural families:

> That time, in five years, more than 20 families left our community – they had nothing left to do here ... Families took out loans from banks, but they inevitably failed to pay. Here in the community, those who had more money started to grow fruits, tobacco, or entered in pig rearing, poultry or dairy [for industries]. Then, we heard [from the neighbors] that some people from here were making money at a food fair ...

Together with seven other families, the Menegat family co-founded the Ecologist Farmers Association of Ipê (APEMA). Together with their new customers, the group of farmers negotiated its own rules, including the exclusion of various hybrid crop varieties (and after that, genetically modified organisms). One contribution to this process came from the endeavor of Vilmar in rescuing creole seeds by creating an on-farm seed bank as well as by contributing to the creation of a communitarian Creole Seeds House. Today, he owns an impressive seed collection. His example has led him into leadership positions, including being on the board of the local Rural Workers Union and as the Coordinator of the regional council of the Ecological Farmers Association. In addition, he is a spokesman for the popular struggle against genetically modified organisms.

The seed-farming arrangements described in Box 9.1 are an example of breaking the human/non-human boundary involved in ecological farming; going beyond the mere characterization of "best farming practices." In reconstructing the value of creole seeds, fresh inter-subjectivities emerge.[20] For instance, the exclusion of hybrid crops was connected with protests against transnational companies and genetic erosion, but the practices related to rescuing and using creole seeds have led to new purposes for ecological food, for example, people recognized that creole seeds added a special quality to food. In addition, ecological food also has constructed a commitment to memory and tradition related to creole seeds. Various practices were reconfigured and new relations emerged through the interaction of people, artifacts, organisms, and objects (Schatzki, 2002), changing the material and immaterial frames of daily life. Every Saturday, customers arrive at the food fair to rediscover a pumpkin that they had not seen since childhood or to discover a new red-spotted field bean, different (or lost) tastes and so on.

Box 9.2 Women and agroecology, lessons from Zona da Mata[21]

> To change society the way we want, we take part without fear of being a woman.
>
> *(Women farmers singing before starting a meeting)*

As the sun breaks over the hills and while others sleep, a woman in the Zona Mata region starts her day. A typical day consists of harsh and seemingly endless work: cutting sugar cane stalks, harvesting coffee, cultivating a field under the searing sun, and taking care of the house and feeding her family. A woman's "struggle," as they say, is a fight for recognition, for respect, and for greater space and freedom. Nevertheless, the women of Zona Mata are no longer isolated and alone.

Their process of organization started in the 1980s and was supported by the Catholic Church. A decade later, they founded the Regional Women's Commission, encouraged by regional unions and the CTA-ZM. Their first achievement was to ensure their rural workers' labor rights, as guaranteed by Brazil's 1988 Constitution. This legal recognition provided them with retirement pensions and maternity leave, as well as other basic benefits granted to urban workers since the 1940s. In the 2000s the women started to seek gender rights, including equitable division of labor between men and women and protection against domestic violence.

Along with these struggles, they called attention to women's food produced in home gardens as not only a strategic contribution to food security and family nutrition, but also for empowering women's economic activities. They deepened their knowledge of agroecology and innovated in food processing, marketing and trade, applying new productive skills and strengthening their cause.

Through their organization, the women have forged a dual cause of feminism and agroecology. As a participant explained, "a woman conceives life, which leads her to wanting to 'plant' life. How could she possibly use pesticides in food production, knowing that someone might get sick? ... There is no way to build agroecology while [there is] gender inequality." Through the activities of the Regional Women's Commission, women's rights and gender issues became an important component of the agroecological network as well as of national and international women's movements, intensifying the experience of grassroots groups, while simultaneously strengthening and multiplying linkages with concurrent initiatives located elsewhere.

Box 9.2 summarizes how Zona Mata's Regional Women's Commission engaged with agroecology as a means of emancipation. The emerging associations between women's struggles and agroecology enabled both discursive and material corporealities that influenced the network as a whole, interweaving daily food practices and feminism. This example shows that rooted activity in food gave birth to simultaneous developments in socially intensive *and* politically extensive realities.

These examples – farming, fairs, cooking, singing together – various "sayings and doings" are compounding nexuses, practices that are inexorably bound with materials, such as seeds, gardens and food, creating multiple associations. Evidently, all these interactions emerge from the networks, not ignoring their "transient state" or discontinuities. Nevertheless, it is from the imminent inter-subjectivity that one can perceive ability to cope, for example, with the standardized food market, dependency on hybrid seeds or long-term oppression of women. At this point we bring back the idea of "mutational forms" from Arce and Long (2000), considering that the result of these multiple nexuses are never fully controllable, as rearrangements can provoke other forms of identity, quality, organizational shape, agencies and meanings.

The question, as posed in the introduction to this book, is: why is there so much resistance to agroecology, if it is not really a threat to the conventional agri-food mainstream? One response is that food is literally about life or death, and it holds tremendous political potential: beyond being merely an *object*; food is also a *subject*. When taking into account the various initiatives of agroecology movements, the commitment to producing food according to principles of ecology, ethics, and equity is somehow radical, but moreover, it may deeply shake understandings and feelings of people about life itself. Therein lies a dispute that is discursive and symbolic, but above all, visceral and material. In reference to the Brazilian case, it is also territorial, involving a violent "tug of war" for access to land, but also control over agri-biodiversity, and perhaps life itself.

Following Blackman *et al.* (2008: 14), subjectivity not only serves as a way to understand and tackle neoliberal power relations and inequalities, it also paradoxically reinforces them. Accordingly, we will highlight two relevant concerns. First, the potential subjectivities from associations emerging from agroecological networks cannot be viewed as convergent into a single alternative for conventional food production, otherwise leading to reifications of the dichotomies that we wish to overcome in critical social studies. This leads to an empirically anchored second aspect: a myriad of changing courses rise from agroecological experiences, but the resulting pathways are as heterogeneous as they are unpredictable. Partially because the changes are linked to the growth of (ecological) food as a social phenomenon, a network's expansion is associated with an entanglement of practices, each involving its own socio-material arrangements (Schatzki, 2015) that reinforce the cosmopolitical reach of multiplicities associated with food (Arce, Sherwood, and Paredes, 2015).

Final remarks

The networks that we studied involve individuals (e.g., farmers, technicians, students, researchers and food clients) as well as organizations (cooperatives, trade unions, agro-industries, certification bodies, and state agencies). Over time, these actors gave birth to new associations, assemblages, and bodies (e.g., creole seeds, labels, and ecological food). In terms of objectivities, they have a certain distributive character, for instance, in considering the connections they produced to ensure access to previously unavailable markets (in the Serra Gaúcha) or achieving the establishment of tenant farmers' rights and creating collective strategies for land purchase (in the Zona da Mata). By no means, however, was each of these outcomes the result of a straightforward, instrumental process.

Without denying such attributes from networking, the contingent nature of social practices and material arrangements that emerged through evolving associations was remarkable, as we showed in two vignettes about creole seeds and women's organization. Such associations generated a profuseness of subjectivities, which over time opened up the possibility for multiple courses of change. We would argue that the "diffraction" of affect–subjectivity–affect is, in itself, a product of the expansion of agroecological networks, while being clear about their mutational and transient character. In terms of their ability to creatively generate new food relationships vis-á-vis the dominant mode of industrial food production, ultimately, the agroecological networks are mightier for what they actually do, rather than for what they appear to be or represent.

The main argument developed here is that the heterogeneous sets of associations that constitute these networks are not a direct, instrumental outcome of public policy or a strategic expert or technocratic intervention. At the same time, change in agriculture and food is not merely a matter of randomized or sporadic social processes. Instead, we observed that change had derived from relationships established through situated practices and political activism in public arenas, however unpredictable or difficult to guide or control. The experiences presented here highlight the multiplicity of materials and socials and their endless potential for creative, transformative relational flows. An empirical, nuanced treatment of social networks in food and agriculture involves an effort to go beyond the notion of interactions among singular, autonomous actors (e.g., the market and state) and the narrow route of normativeness.

Notes

1 See Petersen, Mussoi and Dal Soglio (2013), Flexor and Grisa (2016).
2 Throughout this text, we use the locally known Portuguese forms of acronyms.
3 Notably "networks" and "networking" are words deeply incorporated into the social discourse of development as a way to achieve political ends or as "a political value in itself, often in opposition to more traditional (i.e., centralized and hierarchical) organizational models" (Mische, 2003: 2).
4 Refers to a seminal concept in Actor Network Theory. Callon (1986) defines translation as a socio-technical process or organization in which the social and natural worlds progressively take form.
5 Current Brazilian official rules recognize these units as "family agro-industries."
6 The expression in italics corresponds to a category used by local people.
7 In 2015 this network encompassed 329 farming units, four cooperatives, 25 associations, and 30 small processing units supplying fresh fruits and vegetables, beans, corn, grape juice, tomato sauce and *erva mate*, among other products, to different types of markets.
8 In 1991, the Vacaria Project became a non-for-profit organization, the Ipê Ecological Agriculture Center (CAE-Ipê). The current name of the organization, adopted in 1997, is Ecological Center (*Centro Ecológico*).
9 In Rio Grande do Sul a law establishing procedures to control the use of pesticides was approved in 1982; in 1989, regulations were imposed at the federal level. However, these regulatory mechanisms are constantly under pressure, and remain a matter of dispute with agribusiness sectors.
10 Francis Chaboussou, *Plantas doentes pelo uso de agrotóxicos: a teoria da trofobiose*, Porto Alegre: L&PM, 1987.
11 An organization linked to the Catholic Church.
12 See Charão-Marques and Oliveira (2016).
13 Founded in 1998, *Ecovida* includes about 2,500 families of farmers (organized in 213 groups distributed in the southern region of Brazil), 85 advisory bodies, 21 consumer organizations, and 113 ecological food processing units (Perez-Casarino, 2012).
14 This period coincided with the Brazilian transition from a civil-military dictatorship to democracy and the revitalization of civil society.
15 *Fazendeiros* used to control access of people to markets, public services, health care and all sorts of needs.
16 Catholic organizations inspired by Liberation Theology, as in other parts of Brazil and Latin America that created space for re-elaboration of everyday experience and construction of references for collective action (Sader, 1988).

17 At that moment, the PTA project was working to create centers of alternative technology in different regions.
18 Between 1989 and 2010, 161 families acquired land, totaling approximately 702 hectares (Campos, 2014).
19 Colony (*colônia*) corresponds to regions that received European immigrants, but it is used, mainly in southern Brazil, to mean a set of values and identity.
20 Here, we acknowledge the influence of Blanco, Arce, and Fischer (2015) in terms of recognizing alliances between humans and non-humans as part of emergent associations.
21 The video *Outras Marias* (CTA-ZM, ANA and *Formosa Filmes*) also inspired the content of this box. Available at: www.youtube.com/watch?v=3WHzQDIaY44 (accessed 23 April 2016).

References

Arce, A. and Long, N. (2000). Reconfiguring modernity and development from an anthropological perspective. In A. Arce and N. Long (eds), *Anthropology, Development and Modernities. Exploring Discourses, Counter-Tendencies and Violence*, 1st ed. London: Routledge, 1–31.

Arce, A., Sherwood, S. and Paredes, M. (2015). Repositioning food sovereignty: between Ecuadorean nationalist and cosmopolitan politics. In A. Trauger (ed.), *Food Sovereignty in Geographical Context: Discourse, Politics and Practice in Place*, 1st ed. London: Routledge, 125–144.

Blackman, L., Cromby, J., Hook, D., Papadoulos, D. and Walkerdine, V. (2008). Creating subjectivities. *Subjectivity*, 22, 1–27.

Blanco, G., Arce, A. and Fisher, E. (2015). Becoming a region, becoming global, becoming imperceptible: territorialising salmon in Chilean Patagonia. *Journal of Rural Studies*, 42, 179–190.

Callon, M. (1986). Some elements of a sociology of translation: domestication of the scallops and the fishermen of St Brieuc Bay. In J. Law (ed.), *Power, Action and Belief: A New Sociology of Knowledge?*, 1st ed. London: Routledge, 196–223.

Campos, A. P. T. (2014). *Redes sociais, organizações e políticas públicas: interfaces na agricultura do município de Araponga, Zona da Mata de Minas Gerais.* PhD thesis, Universidade Federal Rural do Rio de Janeiro, Brazil.

Charão-Marques, F. and Oliveira, D. (2016). Agricultura ecológica al sur de Brasil: de alternativa a contratendencia. *Íconos: Revista de Ciencias Sociales*, 54, 87–106.

Cintrão, R. (1996). *ONGs, tecnologias alternativas e representação política do campesinato – uma análise da relação entre o centro de tecnologias alternativas e os sindicatos de trabalhadores rurais na zona da Mata.* MSc thesis, Universidade Federal Rural do Rio de Janeiro, Brazil.

Flexor, G. and Grisa, C. (2016). Contention, ideas, and rules: the institutionalization of family farm policy in Brazil. *Canadian Journal of Latin American and Caribbean Studies*, 41, 1–15.

Lanna, A. L. D. (1985). *A transformação do trabalho: a passagem para o trabalho livre na Zona da Mata mineira, 1970–1920.* MSc thesis, Universidade Estadual de Campinas, Brazil.

Law, J. (1994). *Organizing Modernity.* Oxford: Blackwell.

Law, J. and Mol, A. (1995). Notes on materiality and sociality. *The Sociological Review*, 43, 274–294.

Law, J. and Mol, A. (2011). Veterinary realities: what is foot and mouth disease? *Sociologia Ruralis*, 51(1), 1–19.

Long, N. and Van der Ploeg, J. D. (1994). Heterogeneity, actor and structure: towards a reconstitution of the concept of structure. In D. Booth (ed.), *Rethinking Social Development: Theory, Research and Practice*, 1st ed. Harlow: Longman, 62–90.

Mische, A. (2003). Cross-talk in movements: reconceiving the culture–network link. In M.

Diani and D. McAdam, (ed.), *Social Movements and Networks: Relational Approaches to Collective Action*, 1st ed. Oxford and New York: Oxford University Press, 3–31.

Paredes, M., Sherwood, S. and Arce, A. (2016). La contingencia del cambio social en la agricultura y la alimentación en América Latina. *Íconos: Revista de Ciencias Sociales*, 54, 11–25.

Perez-Casarino, J. (2012). *A construção de mecanismos alternativos de mercados no âmbito da Rede Ecovida de Agroecologia*. PhD thesis, Universidade Federal do Paraná, Brazil.

Petersen, P. F., Mussoi, E. M. and Dal Soglio, F. (2013). Institutionalization of the agroecological approach in Brazil: advances and challenges. *Agroecology and Sustainable Food Systems*, 1(37), 103–114.

Sader, E. (1988). *Quando novos personagens entraram em cena: experiências e lutas dos trabalhadores da Grande São Paulo*, 1970–1980. Rio de Janeiro: Editora Paz e Terra.

Schatzki, T. R. (2002). *The Site of the Social: A Philosophical Account of the Constitution of Social Life and Change*. University Park, PA: Penn State University Press.

Schatzki, T. R. (2015). Spaces of practices and of large social phenomena. *Espace Temps.net*, Travaux, 24 March. Available at: www.espacestemps.net/articles/spaces-of-practices-and-of-large-social-phenomena (accessed 8 April 2016).

Woods, M. (2007). Engaging the global countryside: globalization, hybridity and the reconstitution of rural place. *Progress in Human Geography*, 31(4), 485–507.

10

DEEPENING RELATIONSHIPS THROUGH BIO-INTENSIVE FOOD

AgroSano in Oaxaca

Charlynne Curiel

Introduction

Newly arrived in the city of Oaxaca, Mexico, in Spring 2012, I began looking for direct-purchase farmers' markets. I found a popular weekend market, where growers offered seasonal roots and tubers, leafy vegetables, bananas, apples, honey, coffee and mushrooms as well as tortillas and handmade bread, jams, sauces, and dressings. In addition, the market organized public events related to organic production of food, bio-construction, recycling tips, and sustainability.

By meeting local growers, sellers, and food activists, I gained a new appreciation for how these people were more committed to strengthening relations through their activities rather than seeking profit, efficiency, or productivity. Those commitments were embodied in specific food production and consumption practices named by other scholars as: urban (Sonnino, 2009), community, and sustainable (Feenstra, 2002), community economy (Gross, 2011), or local (Crivits and Paredis, 2013). These terms characterize a type of organization based on public control over food – in the name of good, clean, and local food, emphasizing the spaces, actors, practices, and discourses that distance themselves from the "world/global food regimes" (Friedmann, 1982).

With these practices, the local food circuit constitutes a social site to propose and reclaim ideals that range from individual health to ecological concerns. They question productivity models and imaginaries and create practices for a life based on food autonomy and well-being. Food was at the center of people's lives, and networks for locals and newcomers grew around it.

The constitution of sites to socialize aspirations for healthy and agro-ecological food requires consumers to become active subjects and take greater responsibility for relationships generated through their food practices to shape Oaxaca's food-scape. This type of organization is not necessarily alternative, but part of a tangle of ways of growing and consuming food "differently." In this novel space, production and consumption practices update ways of thinking, acting, organizing and feeling towards and through food as a practice of life.

Drawing on Theodore Schatzki's understanding of practice as "the site of the social," and relational social theory, this chapter explores how people involved in *Agroalimentos Sanos de Oaxaca* (AgroSano) – an emergent rural production society in Oaxaca, Mexico – have come to constitute influential relationships and associations around food. Based on ethnographic research and reflexive practice as a participant in hands-on training workshops, I explore how bio-intensive farming[1] and its social expressions inspire wide changes in people's food production, procurement, and exchange.

While such activity is commonly summarized in progressive food studies literature as part of an opposition or "resistance" movement to "global and conventional food production and consumption," it also can be understood as a generative, subjective force involved in giving birth to locally situated imaginaries and potentialities. I will address the case of AgroSano's founders' participation in a wide range of practices and assemblages such as organically grown food products, seed exchange networks, recycling programs, food fairs, celebrations, bio-constructed housing, and alternative energy. This array of encounters effects and affects people's daily experience, leading them to new practices, ideals, hopes, and aspirations for a needed transition towards healthier, more sustainable living and well-being.

Intensifying relations through food

In agro-food studies, common approaches refer to producer–consumer relations as a space of abstract interactions based on ideas of productivity, efficiency, expansion, and growth. Nevertheless, bio-intensive farming practices stimulate other kinds of relations and show other ideals of "growth." While people are interested and committed to different food practices and habits, they constitute other kinds of relations with humans and nonhumans, deepening or intensifying them through organic farming and responsible consumption. I present the practices and narratives around the production, circulation, and consumption of local food as a new site for participation and creation of social practices in which "traditional knowledge," food tendencies, and public proposals to produce and consume in another way are combined.

Fieldwork consisted of visits to markets, greenhouses, gardens, workshops, and training to learn production techniques and purchase local products. I joined a consumer group and shared an interest in the quality of food and the socialization of information about environmental problems. There was a special interest in understanding how it was for people to devote their lives 24/7 to achieve a new kind of life based on practices and values that some years ago they did not even know about, and how they satisfy their demand for better food. How do they act and perform in order to achieve their ideals and dreams for a better life based on self-production and environmental regeneration?[2] What happens with them as they start to consume narratives, practices, and food products, while paying attention to their health and the environment? These questions point to explaining the processes through which growers and consumers committed to organic production, agro-ecological methodologies, and responsible consumption become active subjects constituting and deepening new kinds of relations.

AgroSano is a rural production society located 15 kilometers from Oaxaca. It promotes non-conventional ways of producing, selling, and eating through bio-intensive techniques and direct food buying and selling. The founders, Lupe Chávez and Freddy Armengol, produce organic food and supplies (compost, seeds, and botanical pesticides), distribute their products to some restaurants, and provide vegetables to the tenate básico, a community supported agriculture project. AgroSano also offers workshops on organic food production and home gardening.

In Oaxaca, AgroSano is involved in different kinds of experiences related to food, environmental issues, and ways of living and existing in the city. In this chapter, I study a group of people, including myself, that supported Lupe and Freddy's enterprise as friends and that took part in their farming training workshops.

Food as a "site of the social"

During fieldwork I met people who some years – or even months ago – had never thought about growing their own food, but were now doing it. These moments do not reflect rational motivation, individual decisions, or the screenplay of a cultural structure (Halkier and Jensen, 2011), but express the constitution of a contingent, social, and heterogeneous body interconnected through practices.

Schatzki (2011: 3) notes that the site of the social consists of a bundle of practices and material arrangements, meaning that human co-existence "inherently transpires as part of such bundles." As he explains elsewhere (Schatzki, 2002: 58), "Practice, furthermore, embodies organizations, which circumscribe the meanings and arrangements set up and otherwise encompassed in the intricate understandings expressed in 'doings and sayings'. ... Meaning is not a matter of difference, abstract schema, or attributional relativity, but a reality laid down in the regimes of activity and intelligibility called 'practices'."

Practice studies shed light on how subjects approach food production and consumption, and it explains food practices involving embodiment, the importance of devices, and the significance of pre-conscious feelings as key elements of creativity. Through ethnographic material, I illustrate how certain practices produce, besides food, social relations, spaces and subjectivities connected to knowledge, ideals, values and desires. In other words, I observe "the social" in the emerging food production and consumption practices that are shaping a subjective force that helps constitute local, diverse, and intense relations between people but also between people and nature.

Those involved in responsible food production and consumption practices carry out a number of activities: take care of their gardens, make compost, recycle waste, make organic repellent, save seeds, and avoid industrialized products, transnational stores, and fast and processed food. These practices are changing people's habits, experiences, and imaginaries, and offering other motivations to produce and eat through their social networks, workshops, and participation in different markets and food events.

From telecommunications to bio-intensive farming

After meeting Lupe and Freddy during a presentation about urban gardening, I paid them my first visit in 2013. Our appointment was in their greenhouse behind Freddy's parents' house. Once there I passed through a patio and a farmyard where his parents kept animals: a few horses, a cow and bull, sheep and some hens. There was a wire mesh dividing the farmyard from the greenhouse. Once I crossed the door to the greenhouse, I was already in Lupe and Freddy's facilities. To the right, a tin sheet covered containers of earthworm production and other kinds of compost. On one side there was a small makeshift laboratory where Lupe and Freddy experimented with isolating and culturing fungi and bacteria as additions to organic soil and pesticides.

A long, tall greenhouse lay to the right, shaped by a rectangular galvanized steel tube structure covered with thick, white plastic. I saw tables where Lupe and Freddy kept their tools, materials, scales and other necessities for intensive raised-bed production of vegetables and herbs. Two of the 25 furrows stretched the length of the greenhouse. Most of the furrows were cultivated with assorted lettuces, tomatoes, hot peppers, and zucchini. There were many seedbeds in different stages of preparation. The edge of each furrow contained herbs and flowers, such as rosemary, basil, and lavender to attract beneficial insects and to repel pests. The scenery (or panorama) inside the greenhouse changed dramatically, I was told, due to the continual rotation of vegetables and crops – a central practice of bio-intensive farming.

Lupe and Freddy each studied electrical engineering at university. They met as colleagues at the offices of Telcel, Mexico's largest telecommunications company, and after a short while they fell in love. After some time they married, and eight years later, Lupe explained, they decided to look for an alternative job or a business in which to invest:

> We had a typical job and stressed routine, but we were not searching for a change of lifestyle, we were looking for a business to invest in, an alternative, just in case we decided to leave our jobs and dedicate ourselves to our own business.

Through some acquaintances they got to know about hydroponics and went to a workshop in Mexico City. Once they were back, they started to grow some leafy vegetables. Still working at Telcel, some relative of Freddy told them about a social program of the Oaxacan government whose purpose was to finance rural small-scale production.

Freddy's parents are "campesinos" and own land in a town called San Sebastián Etla. After some thought, Lupe and Freddy applied to the program. While Lupe left her job and decided to dedicate 100 percent of her time to the new project, Freddy kept his job "just in case something went wrong." As he was watching how the small piece of land behind his parents' house was turned into a greenhouse Lupe was supervising the first tomato production.

As part of the program they got technical training under the conventional form of food production: namely, the use of hybrid seeds, chemical fertilizer, and pesticides. Owing to the good price of tomatoes, Lupe and Freddy earned some money. But the next year an excess of tomatoes in the market depreciated the price, and a large number of greenhouses around the state were abandoned, leaving a sort of "archeology" of the disaster of the rural development public policy.

After learning about all the damage conventional agriculture does to land fertility and natural resources, according to Lupe, they searched for alternatives:

> We bought books, visited other people who are producing organically, and started to look for other options. We began to use earthworms and alternative fertilizers, stopped using agrochemicals and continued with tomato production, but organic … But when you learn more you find out that the organic way cannot be single-crop farming.

Lupe and Freddy found out about Ecology Action in California, a center using a bio-intensive method of food production in Mendocino, California. They met John Jeavons, the main thinker behind the approach, applied for a stipendium and went for six months of training in all phases of the bio-intensive course. That experience brought to Lupe and Freddy two certainties: one, they both wanted to dedicate themselves to food production activities in their greenhouse, and two, they decided not to use industrial agriculture techniques that would "finish with the fertility" of the small piece of land that Freddy's parents had lent them.

Upon returning to Oaxaca in 2010, Freddy and Lupe sought to apply their learning to the crops in the greenhouse and they founded a rural production society, *Agroalimentos Sanos de Oaxaca*. At first, production was for self-consumption, but over time they began to look for markets where they could sell their surplus. In addition, they designed courses to train people in bio-intensive farming. The story of Lupe and Freddy is an example of how important contingencies are to transforming people's lives, and how relevant it is to access knowledge, to build new understandings and to appropriate new procedures that provoke rethinking life commitments.

AgroSano: deep, diverse, intense relations

After a couple of months I received an invitation from Lupe and Freddy to participate in a workshop, "Learning to produce our own food," which would take part during all-day classes at their farm for four consecutive Saturdays. I registered and attended the entire course.

Arriving at the first workshop, I found an eclectic group made up of 23 women and 15 men. About half of the participants were between 25 and 35 years of age and most of the rest, older than 35. Four participants were younger than 25, the youngest being two adolescent boys who accompanied their mother. Of the 38 people, 18 lived in the city, while the rest came from surrounding towns in the Central Valley. Eighteen had completed university studies, while the rest had diverse backgrounds – a therapist, two artists, one salesman, one peasant farmer, and some students, as well as a few people who merely described themselves as employed. Just five of the 38 had access to a garden or a plot of land where they could grow food.[3]

In each session Lupe and Freddy developed different topics to explain the bio-intensive approach to farming. Each topic was introduced with a PowerPoint presentation and video projection showing the activities and effects of industrial food. They emphasized the importance of overcoming the "malaise caused by it," the decline of rural productivity, and a loss of trust in the government and the state.

During the first session Lupe took charge of the introductory presentation. To start with, Lupe asked us to present ourselves and explain our motivation for taking part in the workshop. The group shared a common concern over a modern "food crisis," understood not as a deficit in food availability, but rather a decline in food quality.

After showing a short video about population growth, Lupe insisted that "the future seems complicated, and we are confronting a very serious dimension of the problem," [that's why] learning to grow our own food is a form of "life insurance." As part of her explanation about the "distressing contexts," Lupe insisted that access to information is basic: "rich and poor people in general share an interest in good food. We all want to eat well. The problem is access to information … A long time ago, people knew how to eat, what was healthy, but back then the government did not intervene [with rural development policy] … With the intervention processes we all started to be uninformed and then people do not know how to produce as before."

In addition to the PowerPoint presentations, workshops included a number of activities in which people worked in teams. Each of the four sessions was designed to do something "practical" such as: learning to prepare different kinds of compost, to hoe the land 60 centimeters deep and prepare it for cultivation, to make seedbeds, the transplanting process, to prepare organic pesticides with dead insects, chili, onion and garlic or soap, and to recover seeds and store them safely. These activities are involved in nurturing a plant until it is harvested, and the importance of recovering seeds to build a seed bank.

In each of the sessions there was a pause to eat food prepared by Lupe and Freddy with products from their garden. During that time participants relaxed, chatted and, on many occasions, exchanged points of view, information, and experiences about failed experiments in their gardens – like one young female participant whose first compost rotted because she did not put enough soil to protect it from flies – or successful experiences like one adult female who grew cilantro and parsley plants in cans and had enough for her and to exchange with neighbors for other green leafy vegetables.

The majority of people in the group expressed that AgroSano's workshop was their first experience in non-conventional food production training. Very few had participated in other kind of activities like "healthy cooking" or "urban gardening," and even though they received the information and asked questions related to their knowledge or lack thereof – and the possibilities they had in their houses and gardens to produce food – they arrived at their own conclusions and a community of food practice emerged from the group interactions.

There were moments in which we could not hide our surprise when we discovered a natural process considered "almost magic." For example: that earthworm excrement is a great compost and does not smell bad and we can touch it, we can grow – in three months – potatoes in a sack with soil and compost, we can get rid of insects and plagues with a repellent made of cinnamon, pepper, chili, fungi, or, about "the incredible amount of seeds a plant can give us back."

What I realized is that people were very much surprised by "the natural process," "the possibilities nature give us," and "all the things we can do without poisoning land, air and water." At the end of the four sessions in the AgroSano workshop, participants summarized their experience, expressing concerns beyond the modern "food crisis" as:

> The experience and learning are totally new. I am curious about producing and consuming my own food, transferring this knowledge to more people and questioning what we are eating, because it is bad for our health … If we want to live well and take care of the environment, our only option is to produce our own food.
>
> The workshop helped me to cultivate my mind. The next step is to cultivate the soil. It gave me a new understanding about health issues and all the damage we are doing to ourselves and to our children with all the processed food we eat … also that it is important to give back to the Earth a little bit of what she gives us.
>
> I learned things I had no idea that could be done. Now we must take all this knowledge to the practical level in our gardens. More and more people need to be doing this, because is a good option and a very nice experience to produce our own food, without chemical pesticides and fertilizer that damage our health that much.

It was clear that participants were on their way to discovering new relationships between themselves and food, health, and nature. My own perception was that we participants were more "moved" or affected during hands-on team activities than when listening to talks on "the food crisis and the over-exploitation of the natural resources." People were more interested when they started to discover, for example, the different stages of a plant, from seed to fruit, and the time each stage needs, the importance of the soil and its different qualities, or the function of a particular companion plant. Personal, experiential doing and learning had a greater impact on individuals and the collective than information received from others on the need for developing a "critical perspective" on food issues.

Some weeks later, I interviewed Lupe and Freddy about their goals with these workshops. They replied that they wanted to achieve a "public awareness about the importance of producing food and eating well." They saw the workshop as a site "for doing and reflection and not replication, and to get more and more people to produce their own food. In short, training has become a site to build a material-consciousness. If people know how to produce food, they will become empowered."

The objective of this training site according to Lupe and Freddy was that different actors "start a process of recovering and discovering" organic ways of food production. This was part of a larger process that would help people reach food self-sufficiency. In the dynamics of the workshops, I observed how the bio-intensive methodology – as a praxis – interconnected body and mental activities, tools and organic material and knowledge. These interactions socialized and exchanged techniques and procedures among participants, while helping them express personal commitments to another way of producing and consuming food. The workshops are sites that concentrate people as heterogeneous bodies, involving them through practice in a process of learning. Sharing and feeling affects around new issues would little by little become a force for territorial transformation.

Intensifying relationships through community supported agriculture

Initially, Lupe and Freddy sold vegetables through diverse means. At the beginning of each week they delivered a large quantity of leafy vegetables to different restaurants. In addition, they periodically sold an assortment of greens to friends. Nevertheless, when Lupe and Freddy realized that the price of their vegetables, once cooked or prepared on the plates of those fancy restaurants, increased up to five times more than what they were paid, they started to ask themselves who was eating their food. After further thought, they decided to focus on their friends and acquaintances – *la banda* or crew that had greater appreciation for their sweat and tears. At the end of 2014, they started a project called "El tenate básico," inspired by community supported agriculture (CSA).[4]

Freddy and Lupe started with a list of ten people. They sent us an email inviting us to a meeting to share something "new," which was not "the sale" of their products, but asking us to "support" a new "community food production project." The meeting took place in Calabacitas Tiernas, a popular vegetarian restaurant.

In that gathering, Melissa, the owner of Calabacitas, Lupe, Freddy and Clara, a French volunteer at AgroSano, prepared a buffet with products from the greenhouse. They invited us to try the products as a way of introducing their new project. Almost all of the participants present that day knew one another, but our common link was a history with Lupe and Freddy. Among the guests were many friends who had been buying AgroSano's products for a long time.

It was explained that the name of the Project was a joke to make fun of the government's *canasta básica* or "basic food basket" for a typical family. The "tenate" they argued, "sounds more 'Oaxacan'," [5] besides, they added, "This indeed must be basic," in reference to the content (clean, fresh and seasonal products). During the meeting Lupe and Freddy insisted on the terms social "commitment" and "common understanding" to address the kind of relationship we were establishing – they as growers and we as supporters of their project. Paying in advance without knowing the content of the Tenate turns out to be a relation in which "we accept a shared risk" in production, but also a commitment to invest in the farm's diversity. We were asked to be understanding of "the production cycles and the availability of products because they will change by season and natural issues" (such as insect pests and diseases) inherent to organic food production. After reaching consensus on the details of payment and delivery, the ten participants agreed on the "commitment to support" the project. Afterwards we sat around a large table to chat and share fresh salads, couscous, squash hummus, and chard tamales with tomato sauce.

After the first month of Tenate deliveries, Lupe and Freddy called together the participants to "share our impressions" and a plate made with the Tenate products. Nine of the original participants arrived as well as a new person. The loss of a participant opened up our ensuing discussion.

Lupe told us that after the first Tenate delivery, one participant (a family man in his forties), complained about the small amount of vegetables in the Tenate, leading to his dissatisfaction and departure from the group. This situation led to a discussion among the "Tenateros," as we started to call ourselves. Most of the group assumed that the guy who left "converted the [value of the] vegetables of the Tenate into money," which, they argued, was "a big mistake." The price, they said, "Is symbolic," because "the relationships and work involved in [producing] those products cannot be paid." One member insisted: "We are supporting Lupe and Freddy's project because we know them. We know about their knowledge and commitment to food, and because their products are of the best quality."

Others agreed that this project was not "about [establishing] the equivalence of food with money, because those products have values that are non-monetary." Weekly, the Tenate costs 150 pesos (about US$ 8.00). Participants recognized that the same amount of money could purchase a greater quantity of vegetables in a supermarket, but "This is not about quantity; it is about investing in our bodies, health, and contributing a little to the wellbeing of the environment ... What we are paying for is specialized, clean, manual, and careful work, and besides, we know who they are [the growers] and how they produce." Afterwards, each of us left this second encounter with the impression, "We are a community supporting the project of friends who grow healthy and good food."

With the exception of one member who had to leave the group due to time issues, the rest of us continue to be regular participants. At one point, there was a pause in the delivery because of a problem with the greenhouse, but after some months we all got our Tenate again. Sometimes during the weekly pickup, participants meet one another, which becomes a moment to chat and, in my case, to share recipes or cooking tips.

During this time Lupe and Freddy never referred to us as customer or client. This seems to be a special consideration to the relationship participants have with the growers, who are not mere vegetable providers, but also familiar faces and even friends. In this manner the ways we relate socially through the Tenate are completely different from a conventional producer–consumer relation. This experience made me realize that as new or other food production and consumption practices emerge, the social – as local, diverse and intense relations between people and between people and food – is transformed.

The impact of this project may appear small and insignificant from outside the group, but for us participants the experience of being involved with the Tenate represents a substantial departure from the anonymity of conventional food to having a degree of ownership and shared responsibility over one's own food co-production. The first two meetings allowed me to better understand the multiple relationships involved in food, but it also provided me access to a series of new empowering friendships, ways of eating, and economies. Today, a good portion of my food goes beyond simple relationships of commodification and cost per weight and anonymous market exchanges. I am able to stake a claim in my own emerging regime of food, however small in size, that generates health, equity, and sustainability.

Conclusion: from bio-intensive practices to intense social relations

Lupe and Freddy's project shows the potential of practice both as the "temporally unfolding and spatially dispersed nexus of doings and sayings" and as "performance" (Schatzki, 1996: 89). The narratives and activities included in the agro-ecological and bio-intensive techniques performed and embodied by them in workshops and the *tenate básico* consumer group link understandings, principles, values, and "teleoaffective structures embracing ends, projects, tasks, purposes, beliefs, emotions and moods" (Schatzki, 1996: 89). All these elements coordinate practices and imply that practices are tacit and discursive processes, covering mental and bodily processes, taking care of material things and immaterial processes.

Through the intimacy of the workshops, "Learning to produce our own food," and the constitution of the Tenate básico as a CSA project, Lupe and Freddy involved people in new ways of relating to and organizing around food. Their narratives insisted that the process of appropriation of different food practices is viable only when "people see and then believe that it is possible to grow their own food." For them "building trust" and "being a healthy and concrete example" are keys to "transmit the relevance of the labor" they do. These kinds of doings and sayings embodied a regime of activity and intelligibility, where meaning and making sense reside (Schatzki, 2002).

Even though Lupe and Freddy did not look for a change of lifestyle when they left their jobs at Telcel, they opened a new pathway, changing their lives and others' lives for good. Participants in the embodiment and proliferation of healthy food became part of a narrative coming from a political conjunction that reveals the transformation of other ways of constituting everyday existence and the significance of intensifying relations among people. This includes the relationships between people and food as well as other creative alternatives available in the urban space. I observed during the workshop sessions and the meetings of the consumer group that it is through "doings and sayings" that a feeling of co-existence is materially constituted.

The site of training is a "seedbed" of relationships and processes growing a creative scope of new food practices. Bringing food production and consumption to the center of people's lives diversifies the spaces and practices through which people relate to each other, interacting and exchanging ideals, narratives, materialities and dreams. This chapter helps to explain how in this complicated – ecological, political, economical – "momentum," what is at stake is the possibility to experience and constitute other means of production, procurement, and eating, and ultimately, to illustrate the possibilities people have of exploring and finding other ways of living. The bio-intensive methodology applied by Lupe and Freddy in their garden and taught in their workshops can also serve as a metaphor to think of their project as a catalyst of new relations between the social, the natural (the soil, seeds, plants, insects, water) and the material (a greenhouse, organic vegetables, compost, food baskets), bringing a heterogeneity of people together through growing, purchasing, and consuming food, allowing the social to grow and intensify with the potentials of an urban environment.

Notes

1 According to the organization, Ecology Action, the bio-intensive method implies "scientific principles that underlie millennia-old traditional farming systems." Those principles favor planting as many crops as possible in small spaces, using organic supplies and reducing environmental impact (www.growbiointensive.org/grow_main.html).

2 Among people involved in the organic and fresh food movement in Oaxaca, the word sustainability has been criticized, due to the influence of Jairo Restrepo (a Colombian–Brazilian activist and promoter of the organic regenerative production method).

3 I got this information through a survey I carried out the last day of the workshop (October, 2013).

4 This approach has a long history in North America (Hinrichs, 2000; Gross, 2011) and with its particularities is spreading in Latin America (Blanco *et al.*, this volume).

5 Tenate (from the nahuátl "Tanátl") is a cylindrical form of basket made with palm tree leaves. It is used to carry things in the countryside. People carry the tenate with a strap they fasten on their forehead, a method not used in the cities.

References

Crivits, M. and Paredis, E. (2013). Designing an explanatory practice framework: local food systems as a case. *Journal of Consumer Culture*, 13, 306–336.

Feenstra, G. (2002). Creating spaces for sustainable food systems: lessons from the field. *Agriculture and Human Values*, 19, 99–106.
Friedmann, H. (1982). The political economy of food: the rise and fall of the post war international food order. *American Journal of Sociology*, 88, (Suppl.), 248–286.
Gross, J. (2011). Constructing a community food economy. *Food and Foodways*, 19, 181–200.
Halkier, B. and Jensen, I. (2011). Methodological challenges in using practice theory in consumption research: examples from a study on handling nutritional contestations of food consumption. *Journal of Consumer Culture*, 11, 101–123.
Hinrichs, C. (2000). Embeddedness and local food systems: note on two types of direct agricultural market. *Journal of Rural Studies*, 16, 295–303.
Schatzki, T. (1996). *Social Practices: A Wittgensteinian Approach to Human Activity and the Social*, Cambridge: Cambridge University Press.
Schatzki, T. (2002). *The Site of the Social: A Philosophical Account of the Constitution of Social Life and Change*. State College, PA: Pennsylvania State University Press.
Schatzki, T. (2011). Where the action is (on large social phenomena such as socio-technical regimes). *Working Paper 1*, Sustainable Practices Research Group. Available at www.sprg.ac.uk/uploads/schatzki-wp1.pdf (accessed 15 March 2016).
Sonnino, R. (2009). Feeding the city: towards a new research and planning agenda. *International Planning Studies*, 14, 425–435.

11

PUBLIC GOOD

Wheat assemblages and the revalorization of culinary and handicraft practice in Bio-Bio, Chile

Paola Silva, Maruja Cortés Belmar, and Alberto Arce[1]

Introduction

This chapter approaches the landscape as a methodological reality from which to explore the livelihoods of smallholder family farmers, identifying the material characteristics of a territory for a mode of rural life, and analyzing the transformative features of agriculture as a form of existence (see Gastó and Gálvez, 2014; Dewsbury, 2015). The aim of the chapter is to address two public concerns over agricultural realities. First, externalities associated with policies that promote intensive use of the landscape (e.g., the effects of forestation policies) and second, the potential benefits of public recognition in a territory (e.g., the valuation of skills and knowledge).[2] In other words, it examines the material goods and services generated and brought forth in a rural landscape, in this case those related to cuisine and handicraft.

The beneficial effects of agricultural practices are generically understood in this chapter as the constitutive basis for considering territorial knowledge, skills, and services as a public good.[3] The public is understood as a socio-material assemblage, a configuration of practices, forms and bodies that help to make sense of how things, such as wheat, acquire participatory capacity (i.e., agency) in a territory. Food knowledge and handicraft draw our attention to the potential of materials constituting rural forms of existence (Ingold, 2000). They are associative entities, where a social field marked by fluidity and shifting power relations is generating interactions between people and their natural resources. These are material prerequisites for quality attributes and territorial recognition, for stimulating human–nonhuman alliances and for intersubjectivities[4] between people and in this case wheat and its practical uses in everyday life. This is a corporeal and public approach to recognizing social and environmental value, which directs our attention to qualitative entities comprising landscape and territorial formation.

Food knowledge and handicraft are materialities, often overlooked as carriers of potential capacities to achieve equitable distribution and regional development.

In this chapter, landscape is the particular expression (appearance) of a territory; it is the environmental affect and effect of human–nonhuman activity. Thus, a territory is understood as an assemblage interconnecting forms and contents, tendencies and counter-tendencies in a self-organizing and interactive process by which social life becomes a creative process of environmental and social reflectivity, reproducing and transforming specific modes of existence. Underpinning our argument is our concern to advance a revitalization of agriculture and rural livelihoods and to explore the benefits of a social and environmental public value approach. This leads us to ask, what is the significance of recognizing and valuing the interrelations of a landscape, territorial knowledge, skills, and services? How are these interactions unpacked in the notion of public goods (or bads)? In our case, agricultural practices, land use, and crop cultivation comprise an ensemble that we need to explore, via actors' experiences and practices to reassess the significance of commonly used materials in their variegated expressions of organizing practices for effecting change and territorial development.

We are aware that protecting, restoring, or altering the performance of a territory may be part of a broader political and social debate around the natural and social characteristics of a value approach to territorial development. In this chapter, we ask whether such a value and recognition approach to public goods offers a useful ground for identifying the potential of farmers' knowledge, skills, and experiences to choose production and consumption practices freely – in this case, of certain landscape characteristics of agrarian and territorial development. This perspective would seem appropriate for rural policies and projects, which may mobilize, reposition or prevent local actors from using certain practices or taking specific actions that would produce territorial public bads.

Building primarily on an ethnographic approach, information and interview data are presented in ways that privilege local social actors. The aim of the chapter is to shed light on the latent vitality of an "impoverished" territory of little apparent potential (CASEN, 2013). This study is situated in the coastal Itata Valley in Chile (see Figure 11.1), where we are working with farmers to explore the cultivation of wheat. Increased cultivation of wheat for flour production has led to the displacement of traditional wheat varieties used for handicrafts and particular culinary purposes, as well as the related knowledge. The same loss of knowledge has been observed in relation to the the traditional *farro* wheat recipes in Mediterranean areas. In short, food and handicraft are part of a form of existence. The aim of the chapter is to highlight how domesticated, "local" varieties of wheat are sites that encompass public good and landscape properties, which allow the production of handicrafts made from straw and culinary knowledge to exist as part of a regional identity and actors' everyday practice.

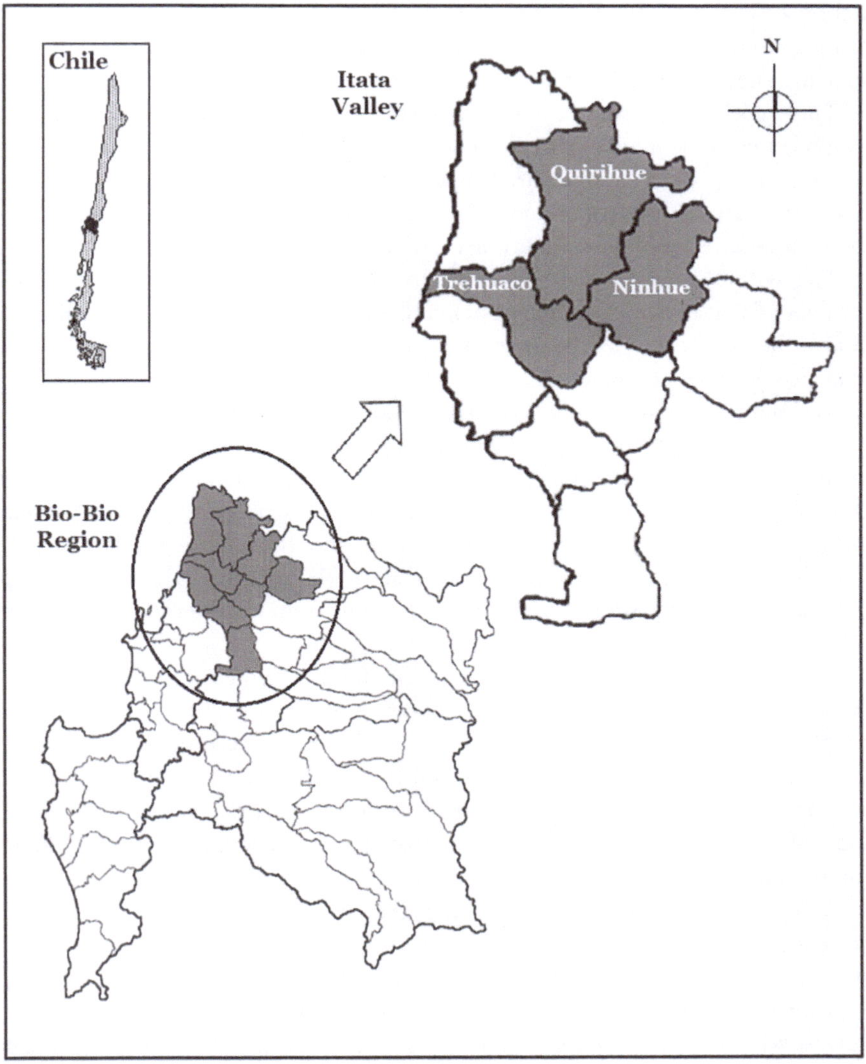

FIGURE 11.1 Map of the Bio-Bio territory

Cuelchas, wheat, landscape and territory

The *cuelcha* is the interweaving of wheat straw through an artisan practice that produces the raw material to make the Chilean *huaso* sunhat, called *chupalla*. Whole families used to perform this artisan technique in the Itata Valley, which is linked to the organic cycle of wheat production within a landscape. The technique involves the selection of flailing straws and the weaving of long strings to make chupallas. This handicraft activity is performed in peasant households, and it is one among a variety of household chores. *Cuelchas* (woven fibre) are constantly in the making around the house, while people talk, prepare a meal or feed chickens and animals. They weave straw several hours a day.

These cuelchas are sold as rolls of fine straw mats, measured according to the length between the middle of one's chest to a fingertip, to hat makers, or *chupalleros*. A completed cuelcha can measure between 90 and 120 meters. Chupalla making is a craft located in the commune of Ninhue, while the cuelchas are produced in the communities of Ninhue, Quirihue, and Trehuaco.

The wheat varieties used by people to make cuelchas are exclusively local varieties of wheat. They are cultivated mainly for their long and flexible stalk and not for the productivity of the grain, as in the case of modern wheat varieties. The Itata people have continued to develop local wheat varieties as supplementary crops alongside the modern varieties they grow for the flour market. Local wheat varieties allow people to practice their craft as predominantly family farmers whose members are constantly looking for monetary opportunities. These varieties are also recognized as making important contributions to what people consider their local food. People still identify wheat varieties by their taste, for example, more or less "sweet," and their texture, that is if they are "soft" or "hard" when the grain is cooked for one of the several local recipes. In contrast, they brand modern varieties as "just good for flour."

The modern, daily bread made from this flour is categorized by people as "weak," extremely white in color and deficient in dietary fibre (bran). Usually, people say that this bread is different from the wholemeal bread they consumed in the "old days." Informants add, "We habitually eat just one piece of bread in the morning to last the whole day. Today, we need to eat three or four pieces of bread and still we are not satisfied." Others say that "the bread today is extremely light and does not fill you."

Although important in the landscape, the area of wheat has decreased 42 percent in the last ten years (ODEPA, 2015). This reduction was similar to that observed at the regional and country levels with 36 percent and 41 percent, respectively. Wheat production in the villages (locally known as *comunas*) under study – Ninhue, Quirihue, and Trehuaco – is well below the regional and national production averages (see Table 11.1). The three villages are located in the eastern sector of the Coastal Mountains (Cordillera de la Costa) in the region of Bio-Bio (36°–37° S) (see Figure 11.1).

Local wheat varieties, in spite of their low yields, are adapted to the natural environmental constraints of the territory, rather than to the country's standardized expected average production. The agricultural techniques used to cultivate this local wheat could generally be described as practices that are no longer part of wheat cultivation in modern agriculture. They are labor-intensive and associated mostly with obtaining the best straw production possible, within the constraints of soil fertility and organic matter content. We observed the coexistence of local varieties together with modern dwarf and high-yield varieties in the same localities, although the latter predominate in the commercially oriented part of the

TABLE 11.1 Average area and wheat production between 1997 and 2007

	1997		*2007*	
Villages	*Area (ha)*	*Yield (kg/ha)*	*Area (ha)*	*Yield (kg/ha)*
Ninhue	1,780	1,220	993	1,203
Quirihue	1,147	1,000	675	1,679
Trehuaco	858	1,040	522	1,260
Total	**3,785**	**1,087**	**2,190**	**1,380**
Bio-Bio Region	105,438	3,280	67,742	4,481
Country	370,684	3,780	219,126	4,922

Source: Agricultural National Census (INE, 2012).

farmlands. The coexistence of different landraces in this biotope[5] is an important feature of the Itata landscape.

In 1938, more than 100 varieties of wheat were cultivated across the country (Mellado, 1997, 2001). As a result of the Green Revolution, the country developed programs aimed at increasing the production of grain for agro-industrial flour and pasta. Local varieties used in the fabric of cuelchas and hat craftsmanship began to be replaced by modern varieties of wheat. However, in the villages in the study, we still found a biotope or pocket of seeds of local varieties of wheat, mainly thanks to the continued practice of wheat weaving that formed a resistance against a complete transition to modern grain production for the food industry.

Local wheat varieties are still farmed because of their wide usage not only in handicraft but also as food. People's culinary knowledge, taste, and familiar characteristics are considerations mentioned by the population as a reason to maintain local foods in spite of the increasing penetration of standardized industrial and global products and considerable physical constraints related to soil and climatic conditions. As a result, the pocket of local wheat varieties was maintained.

In this case, affects and remembering are aspects of food associated with the material dimensions of the landscape. This proximity to the environment and their sense of territorial belonging highlights the significance of actors' experiences, their bodies, and the steady process of imitation and adaptation to the complex plurality of "modern" consumption habits. The influence of cheap and processed food is present in every household we studied; however, people remain close to farming and to their traditional types of wheat. Establishing a territory that is open to outsiders on their own terms and orienting them to experience a sense of control and positioning, local people are socially and economically situated in an unfolding landscape with a performative impact on the generation of public good.

It is the act of being a wheat producer (a knowledgeable and sensitive actor on the properties of specific wheat varieties), a *colchandera* (a person who weaves straw with special skill) or a chupallero (a hat maker) where we find the potential for "wheat assemblage." It is in the connections and relationships between objects, subjects, affects, knowledge, humans, quasi-humans and nonhumans, where these entities generate the potential for the transformation of the territory. A wheat producing territory is assembled and disassembled, involving an active process of politics, affects and laboring to delimit and territorialize different kinds of practices (or expertise). This contributes to locating unwanted public bad (i.e. the forestation of the territory by the timber industry). These realities are either accepted or not within a temporary boundary, and the combination and articulation of social relations may take place, creating a situation from which new territorial entities may emerge as public good.

These realities are forms of political life and experience between social actors and their access to participation with materials and materialities that can be considered as entities that co-produce a territorial public interpretation to deal with problems actors are facing. These explanations cannot be taken for granted as just part of "natural" local adaptation; these are processes involving struggles over access to productive resources, over inputs such as credits, labor, and technology, over opportunities for investment and over the creation of a territory for the pursuit of specific individual and group initiatives.

The vibrancy of the landscape

The climate in this territory is Mediterranean, with an average annual temperature of 14°C, the maximum average of the warmest month (January) of 28.8°C, a minimum average of the coldest month (July) of 3.5°C. The territory has a frost-free period of five months between November and March. The annual rainfall is 1025 mm approx. (ranging between 640 and 1100 mm depending on location) (Novoa and Villaseca, 1989). The dry season runs from November to March (five months) and 80 percent of the rainfall is concentrated between March and August and only 15 percent between September and November (INIA, 2007). Existing water is available for crops until October.

Water, earthquakes, and policies are important entities in the life of the Itata Valley. They are the bloodline of rainfed agriculture, which is practiced by 90 percent of the rural population on a low intensity basis. Plots extend between one and 50 ha, and the main crops are wheat, vines, peas, and lentils. Predominant livestock are sheep and horses, which are fed on natural and artificial grassland where forage species, such as subterranean clover and perennial ryegrass proliferate (INIA, 2004). There are broad similarities in agricultural practices across the three study villages. Although simplified, this expertise – skill – is part of quite a robust landscape, describing a special habitat and broad types of low-intensity rainfed farming that constitutes an energetic farmland-biotope in the Itata Valley.

In this landscape, one can see farmsteads and scattered homesteads with small-sized farms. There is an abundance of pine plantations. The more naturally appearing features in this yellow, green, and brown landscape are the hills and the stretches of varieties of grapes, important to the area alongside varieties of wheat. The names of these varieties include País, Cinsault, and Muscat of Alexandria, and they are used for producing a local wine called *pipeño*. Smallholders grow these grapes using low nutrient and agrochemical inputs because the old vines are well adapted to the rainfed conditions of the area, and they are retained despite low market prices. Don Florencio, a 47-year-old farmer from Threhuaco said, "I resist pulling up my vines, in the hope the grape price will improve." These vines are forms of the corporality of this landscape, where patches of past and present human activity are found side-by-side with the wheat fields.

Our first visit to the Itata Valley for this research was in January and February 2016. When we arrived, knowing the history of the landscape, its wheat, and its actors, our encounter was with a dry, hard, and tough landscape, which made us realize that one of the prominent features of the environment would be the practice of human occupation of a place. Farming here coexists with seismic activity, with earthquakes repeatedly modifying agricultural practices and redefining the landscape.

Andrés Costanzo, a 70-year-old farmer from Quirihue stated that the earthquake in 2010 "changed again how we produce. Before, the water was close to the surface, and you could get water from shallow watercourses. Today the water is quite deep and is no good to drink." Ana Mora, a 40-year-old woman farmer said, "The earthquake has dramatically modified the water courses of the slopes and springs" (Fieldnotes, 2016).

The relation between water availability and soil deterioration is perhaps one of the most important connections people point to as reasons for difficulty in organizing their livelihoods in this landscape. The topography is undulated with slopes between nine percent to 20 percent, limiting the use of land for agricultural activity (Henríquez and Olivares, 1994). This has contributed to soil degradation and erosion. It is estimated that about 83 percent of soils have severe to very severe erosion, mainly due to the high intensity of tillage and fallow land, which leaves the bare soil susceptible to water erosion (INIA, 2004, 2007). These soils generally have poor drainage, low water holding capacity, and low fertility, especially nitrogen (INIA, 2007). The agricultural potential of the territory is marginal at best.

Andrés continued: "We have deteriorated soil. Today we can cultivate the soil only with a lot of care. We still cultivate, but we need to leave the soil to rest between five and ten years to maintain the fertility of the land. The soil needs care if you want to cultivate" (Fieldnotes, 2016).

The Chilean state has promoted a policy strategy to help with the development of the territory: the Decree-Law no. 701 (DL. 701) of forest development which was promulgated in 1974. It intended to encourage afforestation in lands devoid of tree cover and, in many cases, under severe degradation processes. This law, originally conceived for a period of 20 years, established a direct incentive to afforestation and granted tax exemptions, while establishing a number of regulations relating to the management of natural and planted forests (CONAF, 1998).

The direct effects of the DL.701 resulted in the establishment of 1.35 million hectares of pines, of which 800,000 are financially subsidized by the nation-state. This initiative culminated in 1995 with the consolidation of a national forest asset of 1.8 million hectares of planted trees (Sotomayor, Garcia, and Valdebenito, 2002).

The forestry policy incentives and benefits were absorbed mainly by medium and large-scale agro-enterprises, while the small-scale rural sector was left out. This sector managed to reforest three percent between 1974 and 1995 of the total 55,991 hectares of the tree plantations (Cornejo, 1999; Sotomayor, Garcia, and Valdebenito, 2002).

The reforestation initiative was distributed spatially in the regions of Maule, Bio-Bio, and La Araucania. These regions concentrated 77.8 percent of state reforestation subsidies during the period 1974–1995 (CONAF-INFOR, 2004). The forest policy orientation is embodied in the villages of Ninhue, Quirihue, and Trehuaco, where forest plantations cover half the territory of our study. These figures are indicated in Table 11.2.

Nevertheless, the effectiveness of this forest strategy has been questioned. It attached little importance to family farmer livelihoods and produced dramatic changes in the landscape (CONAF, 1998). Low economic benefits for the people inhabiting the afforested territory and risks to the property rights of small family farmers are some of the negative effects of the policy (Chacón, 1978; Fierro and Morales, 1994; CONAF, 1998). Other studies, however, attach great importance to the forestry strategy and consider it a key element in the repositioning of rural areas and the forestry sector in the national economy (Leyton, 2002).

Changes in the landscape have taken place in the past four decades. The inhabitants of the territory have embodied this environmental change in different ways. According to them, the tree plantations have contributed to reducing the number of wheat varieties. Water constraints, and their effects on wheat cultivation, are also attributed to the expansion of the forest plantation, as well as to earthquakes and the state policies for marginal areas. They are a reason for constant lament. People say, "Today, local wells and boreholes are without water." Several people reported that well perforations today are 14 meters deep where in the past water could be found at six or seven meters.

Farmer life-worlds: wheat as handicraft and food

Farmers explain the importance of protecting local wheat varieties by referring to retaining the ability to associate themselves with the material world of the

TABLE 11.2 Forest plantation area in the Ninhue, Quirihue, and Trehuaco reserves (ha)

Criteria	*Ninhue*	*Quirihue*	*Trehuaco*
Forest plantation	11,736	22,350	11,474
Young plantation or just harvested	5,640	12,416	3,908
TOTAL	**17,376**	**34,766**	**15,382**
Area dedicated to forest (%)	43	59	49

Source: Data is based on CONAF (2008) cadastral of native forestland. Figures are adjusted as part of the Orthomosaic Landsat.

landscape: seeds, soil, animals, handicrafts, and food to establish alliances with rainfed agricultural entities. As an entity, wheat is recognizable to local people. They situate and charter their lifeways, assembling their experiences and claims to several life-spheres (Bennett, 2010; Sloterdijk, 2011). Farmers have extended the cognitive to the material, and we argue for an understanding of knowledge as people practice, outside the cognitive and the material duality. This embodiment becomes an issue, not about the body as such but the phenomenon of the "rural body" and the role its practices play within the intersubjective ground of live experience in a territory.

For our case, this is important because the public good and skillful expertise of the peasant body in the Bio-Bio region is oriented to producing good stalks instead of increasing grain production, as suggested in modern agronomy. The peasant body makes the organic materiality of wheat conceivable in the actualization of the regional and distinctive handicraft of the wheat straw. Many farmers still plant small areas of local wheat varieties, either for handicrafts, or to keep their own culinary tradition alive, or because it is the best way to use the steeper slopes. It is the materiality of the local land (or *terruño*), where you cannot use a tractor or a harvester. These farmers prefer handpicking local varieties and they considered these varieties as very healthy. However, they do always treat seeds with fungicides and they control weeds with herbicides. In agricultural actualization practices, they also use chemical fertilizers. Farmer José Sanhueza, 35 years old, from Ninhue said, "I apply liquid once or two times in the cultivation period to control weeds."

Farmers use and adapt new soil conservation practices in order to be able to keep producing local wheat varieties. For instance, this is the case with zero tillage. Florencio Fernández de Trehuaco is experimenting with contour planting and incorporating lupine in his fields as a way to improve his degraded soils (Fieldwork diary, February 2016).

In short, the cuelcha is about unpacking and making visible the colchanderas' practice of weaving, and the continuous adaptation of the cultivation of local wheat. It is mainly women who dedicate themselves to cuelcha weaving, a practice that spans generations, and it is part of a public mode of existence in this landscape. However, this was not always the case. At the beginning of the twentieth century, chupallas were made by women. During the 1940s, with the arrival of sewing machines from America and Germany, the Grossman sewing machine transformed the existing handicraft tradition. During that period, men were in the vanguard of acquiring this commercial apparatus and in their hands a new alliance was forged in which women were displaced from hat craftsmanship or being hat makers. The legitimation of this change was justified by the increase in hat production, and a new organizing nature of knowledge and power took place in the territory. The new relationship between man and machine relocated women into a position where they had to excel in the practice of the art of cuelcha weaving. This was eventually recognized by the chupalleros through the establishment of different monetary prices related to the quality of the cuelchas.

Maria, 65 years old, from the locality of Trehuaco said, "I remember that with my first sale of cuelchas, I bought a potty seat and a teapot for my family." Delfina, 88 years old, from the same locality added, "Cuelchas allowed me to raise my ten children, which is very difficult job for a widow." Eulalia, 60 years old, from Ninhue said, "Each day agriculture and life are more difficult here, what you produce is not worth anything, grapes are thrown away or left to rot on the ground because they have no price in the market. Some years ago, the people who owned a small vineyard would manage well and spent winters without problems, but today grapes don't even pay what you invest in the vineyard, so thanks to the straw, we still weave, and we earn some much needed cash to survive. We cannot exist just as colchanderas, but I can say that the cuelcha provides us with much needed help" (Fieldnotes, 2016).

Locating actors and their practices in relation to territorial matters requires a flexible understanding of the numerous particular relationships that people, their practices and affects with the landscape, machines and natural forces together contribute to maintain "heritage" and to represent a territory. It is people's interaction with local varieties of wheat as the "other" (crops for handicraft), that has generated a local organic alternative to the modern grain market-oriented varieties used to manufacture industrial flour.

Farmers in the region are constantly referring to their culinary heritage, acknowledging that the way they eat is the practice that is helping them to keep growing local varieties of wheat, such as the *colorado* (red) variety. Food heritage is an important dynamic, socially and historically. Ingredients, meals, and cooking are forms of reference to vital and visceral food experiences, which are constantly actualized and deeply affect the involvement of the population with the foodways of wheat, by them typically referred to as "Chilean rice." Delfina said, "The Chilean rice is very nutritious. However, with the arrival of processed food, younger people started to criticize our food, as 'only good for birds,' and they don't recognize our culinary knowledge." Corina Avendano, 35, of Ninhue adds, "to be honest, I am no longer preparing Chilean rice. These are meals for poor and old people. We usually buy rice and pasta" (Fieldnotes, 2016).

Local informants reported that food changes started 55 years ago or one generation back. One example is the school feeding programs. From 1956 onwards, children received reconstituted milk, which they used to drink with homemade bread. The school provided lunch, mainly polenta with salt-dried fish. These institutional food programs were important to complement the meagre local grain diet but partially replaced the family as the unit to nourish the children. The Catholic Church supported the school feeding programs. During the 1960s, each school organized the program according to the food they received from international development donors or the Chilean state. By the 1970s and the 1980s however, the school feeding programs were consigned to private companies. The food started to arrive in tin cans, to be heated and served by the people in charge of the school kitchens.

During the 1980s and the 1990s, local people started to have easy access to processed biscuits and soft drinks from local urban retailers. With time, schools started to use more potatoes and rice in their meals, and children became less used to the consumption of local wheat products. These food changes seem to be associated with changes in health and nutrition within the study area's population.[6]

The identification and reproduction of the different types of wheat are recognized for their weaving qualities, but different wheat types are also savored by actors' food sensitivities. They are described by their performance in the materiality of cooking and in the objectification of the culinary characteristics of the grains. This sensorial and functional identification is a crucial component in the emergence of a public good, overlapping with interweaving discourses, narratives, and interactions of meaning and values of the culinary knowledge around wheat, which cannot be fully grasped without understanding the interactions between humans, nonhumans, and the landscape.

The intersubjective nature of the handicraft and culinary knowledge as a public good has territorialized wheat and people through public classifications of the characteristics of the local wheat varieties and in the unfolding of their material handicraft and culinary properties (see Table 11.3).

One of the people interviewed said that the colorado (red) wheat variety is an indicator for its easiness to be elaborated in toasted flour. First, the grain is crisped in a *Callana* (a metal apparatus to toast grains) and then is ground by hand and mixed with hot water or milk as part of a rural breakfast. This beverage is known as *ulpo* (the term comes from the indigenous Quechua language of the northern part of Chile, and means dry flour). When this porridge is mixed with *chicha* or red wine, it is known as *chupilca*. Local varieties of wheat are retained in the local culinary repertoire because of its ease of being ground for domestic consumption of wholemeal and toasted flour. Local varieties of wheat are also cooked and prepared as whole grain; this is known as *mote* of wheat.

Delfina Neira, 88 years old from Trehuaco said, "Our local variety *fiuto* is the best wheat variety in the world. It is a white, beautiful and a sweet grain. Cooking preparations are usually done with split grain. For *soplillo* flour, the 'protruding' spikelet is harvested green, after which the wheat is rubbed and roasted to remove the peel."

Laura Sanhueza, a 77-year-old woman from Trehuaco, said, "I do not prepare this anymore, but it is delicious because it is sweet and fragrant." *Locro* or Chilean rice is a mature wheat grain, mainly used in casseroles; but also, it is eaten alone or instead of rice. Delfina Neira, 88, from Trehuaco said, "the Chilean rice can replace bread when the meat is cooked, and we have coffee made from wheat." This is prepared from ground and roasted wheat grain (Fieldnotes, 2016).

The culinary knowledge of this farming area represents a public good that could potentially emerge as a small niche food market, offering territorial valorization of the culinary knowledge and sustainable farm food supply. This would add new food products to the industrialized food market and contribute to countering current agro-environmental consumer concerns over food quality and origin.

Today the local wheat varieties and their traditional use for handicraft (cuelchas) and the local culinary knowledge are at risk of disappearing. Young people are

TABLE 11.3 Characteristics of the local wheat varieties identified by rural producers

Name of the local wheat varieties	*Color of the wheat spike*	*Presence of beard*	*Long cycle (winter)*	*Quality for weaving*	*Thickness of the wheat stalk*	*Yield in grain*	*Use: Cuelcha culinary*
Oregon	White	No beard	From May to January	Good	Thin and soft	Good	Cuelcha
Milquinientos	Red	No beard	Similar to Oregon	Acceptable	Thick	Very good	Double purpose, cuelcha, and cooking
Milufen	White	Beard	This is a little quicker than the Oregon	Acceptable but breaks when weaving the cuelcha	Thicker than Oregon and soft	Very good	Cuelcha
Fiuto or Egipcio	White	Beard	Similar to Oregon	Bad	Thicker than Oregon	Very good	Cooking: mote (husked wheat), harina de soplillo (shredded wheat), excellent for soup
Colorado	Red	Beard	Similar to Oregon	Good	Thin and soft	No good	Cuelcha and toasted flour
Chucho	Red	Beard	Similar to Oregon	Soft to weave cuelchas, but the hat does not last because threads cut easily	Very thick	Good	Cooking: mote (husked wheat) and wheat coffee
Onda	White	Beard	Quicker than Oregon	No good for netting	Very thick	Very good	Cooking: Chilean rice
Furfuya	Red	No beard	Similar to Cebolla	Good	Thin and soft	No good	Cuelcha and toasted flour
Cebolla	Red	Beard	Longer time than the quick wheat (20 days longer)	Good	Thin and soft	Good	Cuelcha and toasted flour

Source: Ethnographic study January–February 2016.

migrating to the cities and have lost interest in the countryside. It seems children consider the techniques and uses of wheat as belonging to the previous generation. To address this situation, a negative good, local municipalities have recently started a public program to bring back territorial celebrations, where food and handicrafts are important material entities to maintain territorial heritage and to promote tourism. Finally, the relationship between landscape, vibrancy, and knowledge is affectively expressed in payas, a form of local narrative with its origins in the rural space and in the midst of the rural body:

Florencio Fernández, colchandero and farmer said, "From the land arises the wheat, and with the wheat straw, we made a cuelcha; with the cuelcha a chupalla is made by craftsmen, and these are in Antiquero, Tauco and Pachagua" (January 2016).

Conclusions

In this chapter, we have explored realities of the landscape and territory as a means of understanding the livelihoods of smallholder family farmers in three communities of the Bio-Bio region in southern Chile. The study provides new appreciation of interactive social and environmental relations – a seamless assemblage of the landscape, territory, care, and the public as a right of its human denizens.

Our purpose was to highlight how local varieties of wheat have properties, which in an alliance of practice between local inhabitants and the organic nonhuman entities nourished a site of agency. This is visualized as a composite of an assemblage of multiple and joint sets of material outputs by the agrarian corporality (i.e., the rural body) becoming part of an alliance with the landscape and machines to enhance the potentials of a territory. These happenings are inducing long-term benefits through the maintenance and restoration of wheat biodiversity and territorial development. These characteristics of the act of existence in a territory are allowing the continuation of material activities, such as wheat handicraft and culinary knowledge. These materialities deserve recognition to stimulate territorial potential that must be recognized and valued as a public good.

The recognition of these value-characteristics of rainfed agriculture (known in Bio-Bio as *la agricultura de secano*) and the organic and material resonances are at the core of the social, political, and cultural assemblages of a territory. The value of the landscape and the knowledge of the inhabitants of a specific rural space, presume the manipulation of private and common goods production and consumption activities to reverse trends affecting environmental processes and the generation of public bads. Some of these outputs are the result of historical policies to transform the territory. Therefore, we can draw the conclusion that a valuation of imperceptible materials and materialities must simultaneously consider the landscape and people's practices in its entire territorial involvement and value the associated public performances through actors' experiences of the generated social value. In our case, the visualization of the imperceptible material and materialities of goods and services is achieved by considering a landscape as a territory and analyzing the connections of farmers' lives to handicrafts and food.

Ethnographic research provided a basis for viewing Bio-Bio's territorial development through handicraft and food practices as a socio-political material experience where intersubjectivities and affects revealed a multiplicity of values, as a relationship of territory and public good within emerging processes of agricultural activities and the act of producing, working, and consuming wheat. These imperceptible outputs, which are often uncertain due to the randomness of the forces of nature (i.e., earthquakes) and site (i.e., straw handicrafts), can be considered as agency-specific of potential change (i.e., the maintenance of traditional seed varieties of wheat). With respect to these imperceptible gatherings of materials and experiences of humans and nonhumans, assemblages are likely to be an important focus for future agrofood studies.

"Public good" is not just a concept, but potentially also a methodology or approach useful for identifying territorial potentials. The implication of studying grounded associations, such as those in wheat, is that their intersubjectivities and specificities (practices), multiplicities (alliances between humans and nonhumans) and resulting assemblages fundamentally involve relationships and relationship making, such as those involved in handicraft and food. These associations, assemblages, and relationships are central to the actualities and potentialities of rural life and livelihood. Given their centrality to public development, they merit further attention as a public good and need to be part of a landscape–territorial analysis. Thus, a public good orientation involves a valuation that spans the set of generated effects and affects experienced by actors and their landscape in a territory beyond the simple determinacy and utilitarianism of a "hegemonic" market economic regime.

In this chapter, handicrafts and foods are recognized and valued as non-market-economic goods, produced in the agricultural landscape; they are the organic "other" to modern market grain orientation as a specialized land use practice. This other straw wheat is a public good, a territorial resource commonly overlooked or neglected but of great importance for the revitalization of rural territories.

Notes

1 We would like to thank Eleanor Fisher for useful comments on an earlier version of this chapter. The usual disclaimer applies.

2 Value, the regard that something is held to serve, is conceptualized here as emerging from actions and alliances between human and nonhuman actors. These alliances confer meaning through situated performances that dissolve existing boundaries between the world of subjects and objects. This process of dissolution opens up new courses of action and brings out desire that confers significance on something through the habits of the public and what is ultimately good for material existence or life.

3 A public good is an entity that includes and generates relationships of reciprocity and solidarity that improve the capacity and availability of the general public. In a non-economic sense, the concept is often used to describe something, such as knowledge, skills and experiences, that is useful for everyone. A public good is not limited to economy, which is in contrast to a "common good" that embodies a contested value vis-à-vis private or individual goods.

4 Intersubjectivity refers to alliances that are formed at an interface whereby entities (real or not) construct, reconstruct, and dismantle themselves as they cross each other's boundaries to constitute new individualities, be they linguistic or corporeal, capable of generating a degree of power or potential (Deleuze and Guattari, 1988). Through intersubjectivity, the experience of creativity is available to the public and not just to the individual but to the wheat varieties (nonhuman) other, associating individuality, the social, the public, and the other.

5 A biotope is an area with a distinct set of environmental conditions that supports a particular ecological community of plants and animals.

6 In the three communities under study, we have figures indicating that in Ninhue and Trehuaco there has been an increase in the number of overweight people. The average is considerably higher than the national average for overweight people (see CASEN, 2009).

References

Bennett, J. (2010). *Vibrant Matter: A Political Ecology of Things.* Durham, NC: Duke University Press.

CASEN (2009). *Encuesta de caracterización socioeconómica nacional.* Santiago, Chile: Ministerio de Desarrollo Social.

CASEN (2013). *Encuesta de caracterización socioeconómica nacional.* Santiago, Chile: Ministerio de Desarrollo Social.

Chacón, I. (1978). *Evaluación económica del DL.701 en la VII Región.* Talca: Serie Estudios 1. Facultad de Ciencias Forestales, Universidad de Talca, Chile.

Cornejo, R. (2000). Situación actual y proyección futura del programa de forestación en pequeñas propiedades. In G. Valdebenito and E. Urquieta (eds), *Seminario de investigación y desarrollo forestal en la pequeña propiedad.* Santiago de Chile: Instituto Forestal.

Corporación Nacional Forestal (CONAF) (1998). *Evaluación de resultados de la aplicación del DL.701 de 1974.* Santiago de Chile: Oficina de Estudios de Planificación.

Corporación Nacional Forestal and Instituto Forestal (CONAF-INFOR) (2004). *Exportaciones forestales chilenas.* Santiago de Chile: CONAF-INFOR.

Corporación Nacional Forestal (CONAF) (2008). *Evaluación de resultados.* Santiago de Chile: Oficina de Estudios y Planificación.

Deluze, G. and Guattari, F. (1988). *A Thousand Plateaus: Capitalism and Schizophrenia.* London: Athlone Press.

Dewsbury, J. (2015). Non-representational landscapes and the performative affective forces of habitat: from 'Live' to 'Blank'. *Cultural Geographies,* 22(1), 29–47.

Fierro, G. and Morales, R. (1994). *Evaluación del subsidio a la forestación y propuesta para su continuación después de 1990.* Santiago de Chile: Departamento de Ingeniería Industrial, Universidad de Chile.

Gastó, J. and Gálvez, C. (2014). *Cruz de piedra: limitantes y potencialidades para su desarrollo.* Santiago de Chile: s.n.

Henríquez, M. and Olivares, P. (1994). Condiciones físicas y químicas de los suelos de Quirihue, Ninhue, Portezuelo y Ránquil. *Revista Tiempo y Espacio,* 4.

Ingold, T. (2000). The perception of the environment. In *Essays on Livelihood, Dwelling and Skills.* London and New York: Routledge.

Instituto de Investigaciones Agropecuarias (INIA) (2004). *Manejo y Prácticas Conservacionistas de Suelo para un Desarrollo Sustentable del Secano.* Chillán, Chile: Boletín INIA No 124.

Instituto de Investigaciones Agropecuarias (INIA) (2007). *Cultivo del Cerezo en el Secano Interior de la Región del Bio-Bio.* Chillán, Chile: Boletín INIA No. 163.

Instituto Nacional de Estadisticas (INE) (2012). *Censo 2002 y Proyección de Población.* Santiago de Chile: INE. Available at: www.ine.cl

Leyton, V. (2002). *Informe Final: Evaluación de la Actividad de Plantaciones Forestales en Chile entre 1940 y 2000.* Santiago: s.n.

Mellado, M. (1997). Rendimiento de grano y paja e índice de cosecha de trigos hermanos (*Triticumaestivum* L.) de diferente altura. *Agricultura Técnica*, 2(57), 96–101.

Mellado, M. (2001). Las semillas de un siglo. *Tierra Adentro*, 40, 24–25.

Novoa, R. and Villaseca, S. (1989). *Mapa agroclimático de Chile.* Santiago de Chile: Instituto de Investigación Agropecuaria (INIA), Ministerio de Agricultura.

Oficina de Estudios y Políticas Agrarias (ODEPA) (2015). *Información Nacional de Superficie Sembrada, Producción y Rendimientos anuales.* Santiago de Chile: ODEPA.

Sloterdijk, P. (2011). *Bubbles: Sphere 1. Microspherology. Semiotext(e).* Cambridge, MA: MIT Press.

Sotomayor, A., García, E. and Valdebenito, G. (2002). *Análisis del Programa de Forestación Campesina 1999–2002. Programa de Transferencia Tecnológica para un Desarrollo Forestal Sustentable de Pequeños y Medianos Productores.* Concepción Chile: Instituto Forestal. Proyecto FDI.

12

AFFECTIVITY IN PUBLIC PROCUREMENT

The case of New Dawn Cooperative and the elderly in Argentina

María Laura Viteri

Introduction

The purpose of this research is to explore the affective intensity generated by human–human relationships and human–nonhuman relationships (Archambault, 2016) focusing on a public procurement experience in the city of Mar del Plata (Buenos Aires, Argentina). The many human and nonhuman protagonists of this story include a dairy products factory that has been occupied by its workers and now operates as a cooperative (New Dawn), buildings, elderly adults in various retirement community centers, the Medical Response Program (PAMI, Spanish acronym), and public-sector staff from the Municipal Sub-Secretary of Production, the Ministry of Social Welfare, and other public entities.

Following the flat or rhizomatic ontology that De Landa (1997, 2006) adapts from Deleuze, we attempt to go beyond social categorizations of the subjects that participate in this assemblage of public and private associations. Therefore, we specifically focus on the affective interactions with other humans and/or other types of materialities (e.g., milk, machines, factories and centers for the retired) during the entire public procurement experience. The objective capacities to affect and be affected go beyond linguistic categories, which is the challenge we intend to overcome by analyzing the practical implementation of a policy such as public procurement.

The formation of a consolidated and sustainable "assemblage"[1] could be one of the fundamental conditions for an intervention policy to be a collective construction. The cooperative associates' need to maintain their work positions involved developing solidarity networks with diverse subjects, for example, the members of the Dairy Industry Workers' Association of the Republic of Argentina (ATILRA, Spanish acronym), lawyers with experience in occupied factories, and municipal and national public officials. At the national level, they managed to obtain the support of the Ministry of National Social Development. This ministry declared the machinery installed in the factory as public goods, thus avoiding the loss of working capital in case the company removed material or expropriated it. The ministry offered professional workshops on cooperative management, and also offered a subsidy for a machine that fills sachets with milk.

The interactions between these workers and other actors from the public and private lattices are to a certain extent predicated on previous struggles (neighborhood assemblies, the *piquetero* movement, movements of the unemployed) that enabled the construction of a better-organized civil society capable of defending its rights. The possibility of bringing about policies emerging from the organizational experiences of the territories' own actors is evidence for a certain degree of assent among the public officials who are open to more inclusive political activity. These social assemblages that challenge the development models imposed by various instances of the State government prompt us to rethink intervention practices from a perspective of inter-relationships with emergent properties (new knowledge and synergies, some of which are disruptive, others constructive) that can enrich or impoverish this space of interaction among heterogeneous life-worlds (Long and Long, 1992; Arce 2013).

The analysis of these encounters – from the perspective of the involved social actors, and setting aside preconceptions on "power" that might be exercised by the party that intervenes – allows us to document, deconstruct, and demystify the intervention of "planning experts" in development situations. The lessons that emerge from these encounters open the possibility for each of the actors to envision new repertoires, going beyond the instrumental relationship based on the State's institutional hegemony and its political intervention processes (Arce, 2013).

The affect relationships generated in this public procurement web are experienced and constructed in contrast to historic forms of public intervention. Focusing on affective encounters allows us to analyze the creation of intersubjectivities based on the importance of the "affective." These subjectivities, which emerge from the interaction between humans and nonhumans, have the potential to transform daily life with respect to its commitment toward the material world (Archambault, 2016; Stengers, 2008).

The first part of the chapter employs real-time ethnographic elements, walking through dairy production and distribution to various retirement community centers. Once this dairy path is explained, the next section shows how a public policy emerges from the cooperative's private need to sell and the public officials' need to demonstrate equitable and sovereign policies. Finally, a brief section reflects on the importance of affects and intersubjectivities for transforming the world.

Dairy products, social subjects, trucks, and smiles

At eight o'clock on an April morning in 2015, with my camera in hand, I find myself with another colleague from the National Institute for Agricultural Technology (INTA, Spanish acronym) in the "New Dawn" dairy factory, ready to follow the path of the dairy products. The idea is to leave the production factory in Mar del Plata, and go to the two retirement community centers (autonomous civil associations). These centers operate as collaborative spaces for the Medical Response Program (PAMI) to distribute food to elderly adults who cannot afford to buy food in the commercial market.

There is a mural on the factory wall with images of workers doing different tasks that are carried out in a typical dairy industry. Painted by Omar Sirena in 2013, the mural is a physical testament of the support of a local artist for 70 factory workers who, since May 2012, have been fighting to keep their jobs after their bosses' successive bankruptcies. "Yes, we [the workers] are inside a company that was recovered by its workers, where management passed from private hands to the organized efforts of its previously unpaid workers" (Ruggeri, 2014).

The Vega Family founded the former company, "Amanecer," in 1939, and they sold it in the 1990s. The subsequent owners were unable to cover all of its demands, indebting themselves with the providers of basic primary materials such as milk. The company was on the verge of bankruptcy, which endangered the employees' jobs. In this context, the workers, as advised by the Dairy Industry Workers' Association of the Republic of Argentina (ATILRA), decided to occupy the factory and organize themselves as a cooperative. This process demonstrates the feelings that these workers hold for their workplace, which materialized in their desire not only to save their source of employment, but also preserve the machines, their providers, and their clients.

To enter this factory, which was almost not included in the productive environment of Mar del Plata, is to come into contact with a collective construction that demonstrates the human capacity to confront the difficulties of change. For these workers, daily production of fresh foods is part of their "know-how," which allows them to frequent different spaces of reorganization, thus generating new and creative relationships with other subjects in the social lattice. We reflect on this as we sit there, waiting for Fabio, the president of the cooperative.

The factory has a retail store that offers products developed in the Mar del Plata factory as well as those in the Tandil facility (located 170 km from Mar del Plata). This facility is where the hard cheeses requiring longer maturation periods, such as Gouda, are produced. At 45 percent below the prices of competitors, the retail store offers very attractive prices.

Entering the sales room, a visitor feels a certain atmosphere of success, as materialized in the large volume of registered sales. As Adrián (30 years old) restocks the shelves, he proudly tells us:

> Luckily, the people are with us, they support us, and this fills us with pride. Here, in the factory, a lot is sold; that's because we sell at wholesale prices, and we have not lowered the quality of our products. Here, we make milk, cream, *dulce de leche*, yogurt, desserts, flan, and the *Bandoleros dulce de leche* candies.

> And all I know is how to work. It's true that politicians helped us, and those from the directive commission are the ones who have the most vision and they represent us very well: they know more than we do. But, we are all just a grain of sand in the company. Moreover, this is the work of God.

For Adrián, societal support is of utmost importance for regaining self-esteem in the face of the threat of being another addition to the unemployment lines. For many of the workers, keeping the factory open is the work of "almighty forces." They value their own work and their struggle, but they also wager on "almighty" imaginaries.

Immediately, upon his arrival, the president of the cooperative, with whom we organized the visit, greets and invites us to enter the factory. The first thing we see is the room where liters of milk are packaged. At the moment of our arrival, four obsolete machines are being cleaned, along with a newer one (obtained via a set of subsidies from the Ministry of National Social Development). The passing of time shows in the factory. The establishment is over 75 years old, producing milk since 1939. Little investment can be perceived in the building or machinery infrastructure, and there is no security system reflecting modern technologies.

While Fabio organizes product delivery for the retirement community centers, we move up to the second floor without attracting attention from any form of security. Unlike other food processing factories, there is no strict control on the entry and exit of visitors, thus suggesting a certain libertarian ambiance. We run into two workers who are engrossed in their activities of cutting and packaging fresh farmers' cheese. Upon seeing us, one of them removes his mask and gloves, stops his activity, and greets us:

> If you want, I can show you a bit of how we work here. In this area, we are putting together the cheese wheels, but, to be honest, my specialty is the *dulce de leche* candy. Come on, I'll show you … as I am one of the oldest; I know these machines very well.
>
> At the moment, because of a lack of materials [milk], we are not using this cheese room. For now, we are just making cheese in Tandil. We are having trouble getting organized. Not everyone puts in long hours of work; some still think they are working under a boss. But now, we are all responsible.

This worker (over 50 years old) proudly shows his ability and his affective interaction with the machines ("I know them very well"). His story also manifests high levels of heterogeneity in the subjectivities of his co-workers. To go from being paid workers to being collective owners of a factory entails a new mindset and a change in attitude toward their work activities.

While we were with the worker and the machines that make the *dulce de leche* candy, Fabio informed us that the truck was ready to leave with over 1,500 products, including ricotta cheese, unsalted fresh farmers' cheese, yogurt, and skimmed milk.

Awakening affects: the milk-way between the factory workers and the elderly

We left the factory and continued on the delivery truck's route to the first site where goods would be unloaded. The sun shines throughout the city, tempering the cold April day. The truck parks at the door of the "Hope for the Future" center, where the people in charge of the association are already storing other dry foods (under the umbrella of the format known as "corporate social responsibility"). They have the list of names for the elderly adults who will receive dairy products as part of the Medical Response Program's (PAMI) well-being project.

Inside the truck, the two "New Dawn" cooperative associates assemble boxes containing the five products. Curiously, they do not do this before distribution, which would alter the delivery time. We can see high social commitment in each activity. However, several elements of logistical management are overlooked. While we help with the boxes, the cooperative workers tell us how they coordinated this type of delivery with PAMI:

> We reached an agreement with PAMI to deliver about 4,000 sacks (five products) and receive rapid pay. This way, we could produce and provide products to PAMI well, using the same prices as in our retail store in the factory (with a lower profit margin). The agreement only works as long as they pay us immediately, because we need the money in cash. We pay our providers in cash. It's our way of competing with big businesses and improving our image, because the previous owners stopped paying many of their milk providers.
>
> Municipal officials helped us get a few microcredit loans to settle the first PAMI invoice. The first pay came 45 days later, just as they had said they would. But later, they began delaying ... Today, we've gone 80 or 90 days without being paid, and we've already paid for the goods – we paid for the milk in cash. This threatens the functioning of the cooperative.

The workers recognize the agreement with PAMI as a very meaningful point of departure. However, the delayed payment and the contract's uncertainty in the long run shows, among other issues, the weakness of the informal ties in the management of this public procurement experience. On the one hand, the bureaucracy of State bodies does not allow for reimbursement of paid services in suitable timeframes. For a small factory, such as the cooperative, a delay in payment is an obstacle for purchasing inputs, particularly for the milk that they pay for in cash. On the other hand, the contract between PAMI and the cooperative demonstrates good political intentions in a context of limited legal instruments for strengthening ties in a more formal manner that would allow them to continue beyond political party changes in government.

Once the five products are placed in the delivery sacks, the elderly adults from the retirement community center begin approaching the truck. The commission associates write down who comes for the food. When we speak to them, they verbally attest that they do not maintain a partisan political orientation. Likewise, our visit – seen as coinciding with the "New Dawn" cooperative truck – is perceived as signaling that we are "allied" to the national government of the moment. Some of the people who receive the dairy products state:

> "We like receiving these products; we hope this doesn't end."
> "We are very satisfied to receive products of such good quality."
> "We were told that next month we will no longer receive dairy products, is that true?"
> "Thanks, Ms President. Let's hope this doesn't end."
>
> *(Participant observation, 22 April 2015)*

The words of these adults show happiness on receiving free food, especially dairy products, which have a high maternal connotation that they associate with the female president, rather than with a so-called nutritional program emanating from PAMI officials and their interactions with the cooperative as well as with municipal policy-makers. The delivery of these foods to the retirement community centers also mobilizes other social links between families, neighbors, and/or taxi drivers. One can see various elderly adults with reduced mobility accompanied by their grandchildren. Others come alone or in couples. Not all of them live close to the retirement community center and have to pay taxi services. Still, they confirm that it is worth the money to come because they take home high quality products for "free."

Once we finished delivery to this center, we went to another center called "Give Me Your Friendship and I Will Give You Mine." There, the commission explained how they organize themselves to distribute the products, demonstrating the extent to which they value product quality:

> There are five New Dawn products. None of them has salt because we all have high blood pressure: ricotta, milk, I think also a liter of yogurt, a good chunk of cheese. They arrive once a month. So, we go around telling the people at the center several days in advance. Yes, the people like receiving them.
>
> *(Mirta)*

> I am not on the beneficiary list, but I hear the products are good. However, someone gave me a piece of cheese, which I myself didn't like because it didn't have salt.
>
> *(Carlos)*

> I also don't like them without salt, but I figured out how to make do. I grated [the cheese] and put it in a ham pie. I know I shouldn't, because I have high blood pressure … that's why they give it to us without salt, because the

> majority of us need to eat unsalted. They are two small cheeses. I mix one with jam and the other, so that it wouldn't go bad, I grated it.
>
> *(Patricia)*

> The dairy delivery is only for those above 60 years old who receive a minimum retirement pension, and for those who have some kind of disability. However, it annoys and bothers me that some come to get the sack that is probably worth US$ 130 [Argentine pesos, or US$ 8.6] in pick-up trucks worth US$ 400,000 [Argentine pesos, or US$ 27,000] … this has always happened, huh?
>
> *(Carlos)*

Generally speaking, all of the elderly adults support this dairy distribution policy. Some of them do so from their own experience, despite the fact that they admit to cheating by mixing the cheese with products that are high in salt (e.g. ham). There is a certain degree of disagreement regarding who should receive these products "for free," suggesting disparities between the "needy" and those that have the economic capacity to purchase these products in the market.

Beyond the disparities and multiple perspectives observed during the delivery morning, as a researcher, I found myself wrapped up in a lattice of intersubjectivities built upon interactions between the workers and their machines, affects for their workplace, and the elderly adults' contentment in being made visible by means of a material gift based on the right to health. Following Stengers (2008), here intersubjectivity is understood as the creation of the subjective through social interactions. As such, subjectivity is the production of resistances to previously established moral models; it is not meant to be understood in opposition to the "objective." These intersubjectivities ask us to analyze a social space in function of its capacity for change. The production of subjectivities shows that those who have previously been silenced now generate their own stories and points of view. Among the subjectivities, there is the pride of the workers in meeting the challenge to change their employee mindsets and bodies, and become bosses and owners of their workplace. Likewise, there is the happiness of the elderly adults in accessing dairy products that are associated with the saying "healthy foods that provide calcium for the bones."

Nevertheless, it would seem that the construction of new subjectivities and the social support that the factory workers received from artists, neighbors, politicians, academics, elderly adults, providers, clients, and so on, was not enough to cement the public procurement experience such that it would continue beyond this point in time.

Constructing solidarity-based and affective assemblages

The 2015 public procurement experience between the associates of the New Dawn Cooperative and PAMI officials is not a spontaneous practice, but rather the result of multiple interactions and pursuits that go beyond the involved subjects themselves. Some have the need to increase their sales (the cooperative) while others are trying to make visible their political practices in order to improve the presence of the cooperative into the city's socio-political lattice (Municipal Sub-Secretary of Production) and to experiment with improving elderly adults' food habits (PAMI, retirement community centers). Together, they interlace their multiple subjectivities that affect and are affected (Stengers, 2008).

The experience recounted here demonstrates the importance of the affective commitment of each of the subjects involved. As the Municipal Sub-Secretary of Production suggests:

> Local governments' commitment to assume their roles as defenders of the local production of fresh foods is fundamental. To guarantee the community's food security means to support local producers with the capacity of producing and distributing healthy foods. That is why we supported – to the extent possible – the recovery of the New Dawn factory. This not only involves productive development and local food security, but also the preservation of jobs, and the recovery and defense of the symbolic, cultural and social patrimony of our community.
>
> *(Jorgelina, representing the Municipal Sub-Secretary of Production, 2015)*

Backed by the social support of Mar del Plata inhabitants, cooperative workers are an embodiment of social assistance that any politician (that is, any with social equality objectives) would want to appropriate due to its success. That is how the Municipal Sub-Secretary of Production became involved in the commercial strategies of cooperative associates, offering other viable alternatives such as the farmers' markets and the relationship with PAMI officials.

Supported by a legal framework, the official entity utilized the national tender system (Decree 893/12) that the nation began regulating in 2012. Under the category of "sustainable procurement," this decree allows public entities to contract goods and services using criteria in addition to economic ones, such as considering social impacts for the eradication of poverty, equal distribution of resources, and working conditions of the most vulnerable sectors (Gabutti *et al.* 2014).

The decree and the personal and political relationship between public and private actors were key elements for establishing the first negotiations and the offer for public procurement of skimmed milk products from the cooperative workers. The aims of these interactions are multiple and heterogeneous. The current regional director of PAMI for the Mar del Plata area recounts – in his political and partisan language – how the initiative began:

> From the day that we began managing PAMI-Mar del Plata, we knew that we could improve the articulation with the retirement community centers, which had always received food donations for economically disadvantaged elderly people. We wanted to improve the social aspects; that is, that people

> not only receive foods, but also get to know with whom they are eating. And if they are alone, the aim is to organize more social events within the centers. We convened the Federation of Retirement Community Centers to understand how they organized their food purchases.
>
> That is when we met the woman who was the Sub-Secretary of Production for the General Pueyrredón municipality. She introduced us to a number of persons from family agriculture entrepreneur cooperatives, including Fabio from the Milk Cooperative.
>
> *(Fernando, speech at the International Conference-Workshop "Public Procurement for Family Agriculture," Buenos Aires, 5 May 2015)*

The actions of the officials who recount the inception of this public procurement experience depend on previous organizations. Both the cooperative and the retirement community centers are expressions of an organized civil society that allowed these officials to interweave their political proposals. Meanwhile, the factory workers and the elderly adults were also affected by the launch of public policies that they themselves helped bring about.

In Argentina, adults over 60 years old who have already retired from work are organized in "retirement community centers." These organizations – which function at a neighborhood level – have historically driven social policies, as this is an enabling environment for the development of multiple activities that allow for the socialization of elderly adults. These centers allow elderly adults to feel that they are a part of the community in which they live and are active in tasks useful for themselves or for others. It is a space for encounters in which they talk, play cards, organize trips, and learn crafts, such as weaving, sewing, painting, ceramics, handicrafts and folk dancing. In summary, these centers play an important role for elderly adults to feel closer to the community, to understand their social rights, and to carry out a healthy life project beyond just their families (García Sotelo, 2012).

Across the country, there are over 5,000 retirement community centers, which are subject to receiving subsidies from PAMI to improve aspects related to health and disease prevention (www.pami.org.ar). The population records of Mar del Plata show a high level of elderly adults. Many of them are active in the 98 retirement community centers distributed across the city's neighborhoods. The centers place a high value on the independence of elderly adults, which is understood in terms of self-determination, and is in opposition to the interference of any other organization in their personal decisions (Molinari, 2000). As a result, the individuals interviewed commented that the vision of the directors with respect to PAMI food distribution diverges slightly from that of the officials:

> Food distribution is a complementary element. We are a support center; we have nearly 960 members in this center alone, okay? People come every day to take yoga classes, participate in weaving workshops, or have pedicures, dance classes (salsa, tango), and soon we will have *reiki*. We also play cards, bingo, or shuffleboard. All workshops are free (subsidized by PAMI).

> Members only pay US$ 15 [Argentine pesos, or US$ 1] monthly.
>
> *(Carlos, 7 August 2015)*

> This is why food distribution is only one part of it. Really, it is a PAMI activity, and we are just intermediaries. We give a hand to get a hand. They need to deliver and be in charge of the logistics. But we give them the space and the willingness to give it. Not just in this center, okay? PAMI sends us a list of beneficiaries for the food and we distribute it. I prefer this system over the farmers' market fair.
>
> *(Patricia, 7 August 2015)*

The testimonies collected in the retirement community centers express their nonpartisan orientation from the point of view of their discourse. However, they all highlight the role of PAMI in sustaining the well-being of the country's retired adults. One adult belonging to an opposition political party expresses:

> PAMI is very good, in all-caps. I am thankful to PAMI. I have a good opinion on what PAMI does. The retired are served very well. Argentina is one of the best countries for retired people in the world.
>
> *(This affirmation, made by Carlos, is backed up by all of his peers, 7 August 2015)*

The interviewee is not only speaking about the centers' food distribution service, but also about its medical services. In Argentina, all PAMI members can receive free medicine for age-related chronic conditions. The consolidation of the centers' network allowed the PAMI officials to go beyond the initial farmers' markets for low-cost foods. A PAMI doctor recounted:

> In 2013, at the door of the PAMI headquarters (in Mar del Plata's center), we set up a fair with the intent of bringing together the family-level micro-enterprises' need to sell and the low-income grandparents' need to purchase at low costs. Of course, we had to transport the retired people to the headquarters … On the days of the fair, the fair vendors and the shoppers added to the 1,500 people that already pass through daily.
>
> The fairground was in the street; on fair days, we closed the road to traffic. Over 200 employees work in this headquarters. Can you imagine what the workers said in that moment? They couldn't park, and it was an inconvenience. Okay, so we included them as well, and we explained the objective of the fair. But this fair only brought together supply and demand. We had to build a public space, not a State one.
>
> *(Fernando, speech at the International Conference-Workshop "Public Procurement for Family Agriculture," Buenos Aires, 5 May 2015)*

The construction of a public space that outperforms a State space demonstrates the need for a new way of policy-making. The complex and heterogeneous social lattice demands spaces where different interests and powers can reach clear goals, such as access to healthy and solidarity-based food for all. The fair initiative allowed Fernando, a doctor at PAMI, to construct new interaction networks between elderly adults and the entrepreneurs. In his words:

> At the same time that the process of recovering the milk factory was underway with the support of the Ministry of National Social Development, we found it opportune to contemplate a program that would allow for an increase in the consumption of skimmed dairy products among the elderly adults enrolled in PAMI.
>
> Launching the purchase of milk products for improving elderly adults' food habits required modifying the previously established public procurement regulations. It was not a matter of simply making a decision, but also of purchasing dairy products from a factory on its way to bankruptcy, of negotiating with all of the retirement community centers to set up the logistics for distributing five dairy products to 4,300 enrolled members and retired people in Mar del Plata, and of requesting the support of the PAMI authorities.
>
> *(Fernando at the International Conference-Workshop "Public Procurement for Family Agriculture," Buenos Aires, 5 May 2015)*

Fernando refers to the complexity of coordinating the efforts of different domains, including his own workplace. To contract a cooperative with scarce financial capacity in the context of the State's payment bureaucracy was a challenge that surpassed the political good will of the interested parties. Among the resources available from the Ministry of National Social Development, they sought a microcredit loan that allowed them to cover costs for the first delivery in cash. Doing so, they met a dual objective: to offer high quality products from one of the city's emblematic factories, and to maintain the jobs at a failing factory. According to the interviewed officials, this experience is the fruit of the active efforts of the PAMI technical team, and in particular, of a strong "want-to-do" political conviction.

Nevertheless, the initiative experienced difficulties in the payment chain. This discontinuity affected elderly adults as well as cooperative workers. At the moment, product delivery is suspended. According to the Sub-Secretary of Production, this management space would have persevered if there had been institutional-level structural consolidation:

> The only space we have to exercise our power is in the public sector, when something gets recorded in a regulation. And regulations lacking significant weight are also useless. It's very difficult. You think you've made some progress, and then when you get back to the structure, you haven't made any progress at all. Politicians are very vulnerable to the spaces of disagreement. They receive pressure from different kinds of lobbies, from the judges to the partisan politicians.

> The necessary step is to institutionalize public procurement. There are different levels of disagreement: one is the relationship with the "public" aspect, to do what the big companies do beneath the table … as we can't lobby like they do, we have to work in the public space.
>
> *(Jorgelina, on behalf of the Municipal Sub-Secretary of Production, 7 July 2015)*

This testimony makes explicit the challenges of the State's governance. For this official, relationships with the partisan political power are not experienced equally by businesses with lobbying capacity as by small businesses with low visibility and minimal negotiation tools. It would appear that the latter can only generate power in the public sector, accompanied by the support of society, the neighbors, of those that have known the factory since they were children, and so on. Winning over the public space is a fundamental factor for popular economies to be able to ground themselves in laws and regulations. Without winning over this space, it is very challenging to sustain an abstract law over time.

In summary, the different networks of actors that form this public procurement experience extensively cross the boundaries of class, organizations, and belief. The challenge is to sustain these complex networks of heterogeneous social actors with multiple and opposing interests. Adding to the challenge are the abrupt political changes within a tumultuous Latin America, which has experienced more time under neo-liberal governments than under populist ones.

Final thoughts

Adopting Stengers' (2008) idea of trying to change the world and not simply understand it, all of us who supported the State procurement initiative are fighting for the replication and emergence of this type of experience. We understand that the emergent properties of the interactions analyzed here increase in quantity and quality due to the affective emotions, as well as the identities and subjectivities of the subjects who are affected by, and affect, the assemblage of the public procurement (Thompson, 2007; Fernández Álvarez, 2006; Paredes *et al.*, 2016).

A policy that reconnects subjects in a creative and innovative way generates nonlinear transformations (De Landa, 2006; Beck, 2006) that demand an analysis from a social critical perspective. This work was grounded in ethnographic tools to detect the relationships and alliances developed between the subjects and their affective materialities. The novelty of the analyzed experience is its disposition to embrace the unexpected and to create affective and solidarity-based ties between humans and artifacts. This allows us to observe a creativity that breaks with traditional forms of industrial assemblages or with traditional public procurement processes that are pre-established by lobbies and bureaucratic dogma.

As people working in science, we accept the challenge of trying to explain this assemblage from the perspective of the corporal and discursive practices of some of the protagonists, as well as from their affective interactions with diverse materialities (machines, milk, dairy products, peers, officials, elderly adults, associates, etc.).

We attempt to unpack a public intervention whose origins are rooted in the construction of social networks. We found that these social networks emerged from different and heterogeneous actors involved in socio-political challenges. By focusing on the subjects and their capacities for action and affect, we discover a counter-tendency to a classic State-government intervention.

The consequences of the orthodox adjustment models applied to the Argentine economy precipitated other forms of counter trends, such as self-organization and collective action. Among such experiences are the recovered factories, which are currently a significant social movement within Argentine society. The consolidation of this type of organizational initiative, which resolves issues such as job loss, demonstrates that social changes are unpredictable, surprising, and non-imaginable (Paredes *et al.* 2016). The challenges they face to maintain their roles as self-managed associates administering a cooperative company leads the workers to collaborate with State officials to engender experiences that overcome political intervention traditions. In summary, State officials must understand the subject – the object of policies – as an active implementer of policies and not merely its "beneficiary."

Another assemblage fundamental to this experience was that of the retirement community centers, which opened their spaces to receive and organize the distribution of the most needed products. While PAMI's intentions were to improve food habits by distributing unsalted products, the elderly adults re-assessed the savings represented by those five products each month. The implementation of a policy that aims to change consumption habits should be accompanied by strategies that reassess taste.

In this experience, it is worth highlighting the acting capacity of the organized and interwoven political subjects, who managed to creatively negotiate their collective decisions with State officials while the officials reformulated their way of policy-making with active subjects. However, this case study leaves questions unanswered regarding the sustainability of the experience in the long run. It would seem that the incorporation of ideas such as social and/or solidarity-based economies in certain State programs is insufficient. It is necessary to strengthen public spaces with subjects that have their own identities and that are inserted in a symbolic and institutional world that represents them. The cooperative's experience, its history of struggle to recover jobs, and its ties to the partisan political world give rise to a social economy that not only produces economic utilities but also constructs society (Coraggio, 2011).

The factory workers' dedication is created not only by their affect for their workplace, but also by the recognition of the civil society that supports them economically and symbolically. However, this level of intensity of affect does not translate to the entire social assemblage of Mar del Plata. It may be said that the creation of intersubjectivities generates potential transformations, but those are not enough to maintain public policies that require political will beyond laws or regulations. Nevertheless, the public procurement experience of Mar del Plata should raise questions about future public interventions. It is essential to strengthen the affective encounters that inspire new ways of being and interacting.

Acknowledgments

The motivation for this article dates back to work completed previously with Ana Karol after the "Public Procurement for Family Agriculture" International Conference-Workshop, organized by the National Institute of Agricultural Technology (INTA) though the specific project of the National Program for Sustainable Territorial Development and of Argen-INTA Foundation, entitled "Commercialization and Market Strategies." This event took place on 5–6 May 2015 with the aim of planting the subject in the national public agenda. I am thankful to the idea exchange and suggestions of Alejandra Moreyra (IPAF Pampeano), Mariana Moricz (IPAF AMBA), Ana Karol (IPAF Cuyo), and colleagues from the workshops in Ecuador and the Netherlands. Everyone, especially the protagonists of the public procurement experience, helped me to develop this work.

Note

1 Assemblage is an ontological approach from Deleuze and Guattari (1993) and used by De Landa (1997). The assemblage is like a constellation and it is made up of heterogeneous elements that are organized through material/expressive and territorializing/deterritorializing imaginary lines or axes.

References

Arce, A. (2013). Conocimiento, espacio y actores en la innovación social. In Paz, A., Paz Montoya, M. and Asensio, R. (eds), *Escalando innovaciones rurales*. 1st ed. Lima: Instituto de Estudios Peruanos, 37–86.

Archambault, J. (2016). Taking love seriously in human–plant relations in Mozambique: toward an anthropology of affective encounters. *Cultural Anthropology*, 31(2), 244–271.

Beck, U. (2006). *The Cosmopolitan Vision*. Cambridge: Polity Press.

Coraggio, J.L. (2011). *Economía social y solidaria: el trabajo antes que el capital*. Quito: FLACSO.

De Landa, M. (1997). *A Thousand Years of Nonlinear History*. New York: Swerve Editions,

De Landa, M. (2006). *A New Philosophy of Society: Assemblage Theory and Social Complexity*. New York: Continuum.

Deleuze, G. and Guattari, F. (1993). *A Thousand Plateaus*. Minneapolis, MN: University of Minnesota Press.

Fernández Álvarez, M.I. (2006). *De la supervivencia a la dignidad. Una etnografía de los procesos de recuperación de fábricas de la ciudad de Buenos Aires*. Doctoral Thesis, Faculty of Philosophy and Letters, Universidad de Buenos Aires.

Gabutti, S., Samper, S., Zárate, C., Gamarra, J. and Vottero, L. (2014). Breve análisis del sistema de contrataciones a efectores locales y de la economía social. In *II Congreso Nacional de Administración del INTA*. Buenos Aires.

García Sotelo, V. (2012). La función social de los centros de jubilados. Mendoza online, www.mdzol.com/nota/393313-la-funcion-social-de-los-centros-de-jubilados/ (accessed June 2015).

Long, N. and Long, A. (eds). (1992). *Battlefields of Knowledge: The Interlocking of Theory and Practice in Social Research and Development*. New York, Routledge,

Molinari, J.M. (2000). Cultura institucional de los centros de jubilados: un encuentro entre los campos antropológico y psicológico. In *Congreso Argentino de Antropología Social*, Universidad Nacional de Mar del Plata.

Paredes, M., Sherwood, S. and Arce, A. (2016). La contingencia del cambio social en la agricultura y la alimentación en América Latina. *ICONOS*, 54, 11–25.

Ruggeri, A. (ed.). (2014). *Las empresas recuperadas: autogestión obrera en Argentina y América Latina*. Buenos Aires: Editorial UBA.

Stengers, I. (2008). Experimenting with refrains: subjectivity and challenge of escaping dualism." *Subjectivity* 22, 38–59.

Thompson, E. (2007). *Mind in Life: Biology, Phenomenology, and the Sciences of Mind*. Cambridge, MA: Belknap Press.

13

ASSEMBLING RESPONSIBLE FOOD MARKETS

The case of *Cooperativa La Manzana* in southern Chile

Gustavo Blanco, Jilles van Gastel, and Andrés Lagarrigue

Introduction

This chapter explores the emergence of a movement for responsible food consumption in the city of Valdivia, in southern Chile: *Cooperativa La Manzana*. During the last two decades, citizens' movements and organizations playing a role in the democratization and re-embeddedness of food production and consumption have been extensively studied in Europe and the United States (Harris, 2009). These movements have been conceptualized and grouped under the academic labels of "alternative food networks" (AFNs) (Renting *et al.*, 2003; Maye and Kirwan, 2010; Goodman *et al.*, 2012); "Short Food Supply Chains" (Ilbery and Maye, 2005; Aubry and Kebir, 2013); and, more recently, "Civic Food Networks" (Renting *et al.*, 2012; Anderson *et al.*, 2014). Furthermore, a number of critical analyses about their contribution to rural development and to wider changes in food systems are available (Goodman, 2004; Van der Ploeg and Renting, 2004; Harris, 2009; Parkins and Craig, 2009). We acknowledge the relevance of this literature in documenting these food initiatives in particular situations in the northern hemisphere. We question, however, the universality of its explanatory value for describing and explaining food experiences elsewhere. Although AFNs are described as embedded processes, their understanding relies on a number of binary oppositions – alternative/conventional, global/local, subject/object – that obscure their relational and heterogeneous composition. Moreover, in this literature there is little attention to the importance of affects and experiences in the constitution of such networks. Hence, the relevance of this case study is twofold: first, it documents the repositioning of solidarity regarding food consumption in social sites distant from the northern hemisphere, and second, it exposes the shortcomings of current conceptual apparatuses in theorizing these movements.

We first briefly present the debate on AFNs by considering two questions. Do these frameworks provide insights into processes and experiences in social sites other than Europe and North America? Do its concepts help to understand self-organizing processes that are creating new markets and new subjects? Next, to address some of these theoretical shortcomings and to shed light on issues raised in these questions, we draw on Assemblage Theory. We explore assemblage thinking in order to understand these food initiatives relationally, as part of multiple experiences and arrangements of various entities in constant becoming. In this sense, we describe the processes of formation of the cooperative through agents, organizing practices, events and contingencies, giving way to a vibrant food assemblage. We pay specific attention to the flow of matter-energy, emotions and lived experiences involved in the highly contingent, open-ended processes of constituting a responsible market. The centrality of affects in the process of creating a new subject (the "responsible consumer") and new objects (the food cooperative and fair food) highlights the need to move beyond structural and materialist understandings of an assemblage.

Alternative food networks: opening a debate

The literature on AFNs emerged during the 1990s following the recognition of changing patterns in the relationship between food production and consumption. A number of authors have described the emergence of food networks as a process of differentiation from conventional agri-business food systems or food chains. Some of these differences were built around the construction of quality (Marsden and Arce, 1995); the importance of place and embeddedness (Arce and Marsden, 1993); or the connectedness of distant agencies (Whatmore and Thorne, 1997). Renting *et al.* (2003: 394) define AFNs as "a broad embracing term to cover newly emerging networks of producers, consumers and other actors that embody alternatives to the more standardized industrial mode of food supply." According to Lamine (2005), AFNs represent strategic responses to the uncertainties caused by the successive food crises that have affected the entire global food system. Based on case studies in France, She points out that these consumption networks offer security to producers (in terms of income) and consumers (regarding production practices) (2005: 324). These networks, however, do not necessarily reduce food uncertainty, since the assortment and quantity of products varies depending on many factors, but consumers accept this type of food uncertainty in exchange for food security. Brunori *et al.* (2012) study the Italian Gruppi di acquisto solidale (solidarity purchasing groups) and explore to what extent these AFNs challenge the dominant practices and governance pertaining to the food system. They conclude that food consumers' networks can easily move to other consumption goods and that organizational innovation can turn into institutionalization. Though many authors celebrate the rise of these new food movements, AFNs have also been criticized for being elitist expressions of aesthetic consumption within an unchallenged dominant capitalist food system (Goodman *et al.*, 2012);

or for its potential to promote a counterproductive defensive localism (Winter, 2003).

The literature on AFNs undoubtedly represents a valuable contribution. It describes and documents emergent markets, new forms of transactions, and new identities and actors, such as citizens, around short food supply, thus bringing to the fore experiences often black-boxed as "alternative."

We argue that this literature exhibits three conceptual limitations. In the first place, the theoretical framework is based mainly on the analysis of experiences in northern America and Europe. Yet, there is no basis to support the assumption of a universal food experience and, as a result, the construction of a singular alternative food geography. Rather, a number of interesting experiences in Brazil and Ecuador shed light on heterogeneous origins and compositions. Some of them are being driven by farmers' organizations (Scarabelot, 2012), alliances between consumers and producers (Coelho de Souza, 2008), or a wide range of civic organizations (Arce *et al.*, 2014; Preiss and Marques, 2015). Others are the result of state-led programs for food supply (Sonnino *et al.*, 2014). These cases show the heterogeneous origins and composition of food movements elsewhere. Any similarities and connections as well as their differences between food processes and experiences in different parts of the world cannot be taken for granted but need to be elucidated.

Second, the literature on AFNs tends to focus primarily on relationships between human actors and a limited number of accepted (good) values. This is due partly to the use of dichotomous distinctions expressing marked value laden judgments such as bad–good, conventional–alternative, formal–informal, micro–macro, local–global and society–nature. This conceptual application of dichotomies is one of the most problematic limitations of the AFN approach. These dichotomies are presented as part of rigid contrasting values and practices, leaving little room to observe the heterogeneity and multiplicity involved in the formation and proliferation of those networks. A closer look at these initiatives shows that an "alternative food paradigm" is also composed of contradicting values, modern ideas, capital, merchandizing, state support and a messy, but not necessarily chaotic, range of agents (humans, other-than humans, and things), with some of them located in other-than-local sites (see Introduction, this volume).

Third, the literature stresses the importance of solidarity, identity, and morality in these networks, but it fails to include affects, emotions, and lived experience as constitutive elements of diversifying foodscapes (Carolan, 2015). A crucial question then emerges: what other theoretical possibilities are available to shed light on the presently neglected importance of affects and experiences in the constitution of diverse food realities?

Food assemblages: objects, subjects, and affects

Assemblage thinking offers a relational ontology of the social that enables us not only to overcome the aforementioned dichotomies, but also to open space for multiplicity and indeterminacy (Anderson and McFarlane, 2011). This is not equivalent to denying the social existence of those dichotomous categories, but to paying attention to how they are produced and to the generative possibilities once they are folded into their points of bifurcation. As Anderson and McFarlane (2011: 124) have described it, "assemblages are composed of heterogeneous elements that may be human and non-human, organic and inorganic, technical and natural. In broad terms, assemblage is part of a more general reconstitution of the social that seeks to blur divisions of social–material, near–far and structure–agency." The understanding of the social thus unfolds from the idea of assemblages consisting of heterogeneous components, be they material or expressive, and whose identity and stability depend on movements of territorialization and deterritorialization (De Landa, 2006).

An assemblage is stabilized by movements of territorialization, when through certain socio-spatial processes of identity formation and homogenization, the degree of internal cohesion of its components increases. Conversely, assemblages are destabilized through deterritorialization, namely, when its components and properties are modified and, eventually, transformed into another assemblage. Territorialization and deterritorialization are not a new pair of dichotomous concepts but heuristic denominations that contribute dialectically to materialize and embed certain associative processes, thus creating new entities, for example, a public policy, an organization or new food objects and subjects. In turn, these movements also dissociate, transform, or re-signify entities that seem stable and widely accepted as well as processes that appear irreversible and consolidated. For instance, the movements of deterritorialization are important to understand and explain the process of contestation of industrial and fast food that until recently seemed unchallenged, thus, opening space for other agents to territorialize healthy and responsible food consumption.

In the light of these ideas, we follow De Landa's understanding of markets as self-organized assemblages of heterogeneous groups of humans wherein complementary economic needs enmesh (De Landa, 1997: 31). Markets, ranging from local fairs to national-level wholesale markets, are understood as being composed of contingent but necessary relations between parts and wholes, including structures, objects, agents and codes (text, genes, and language) (De Landa, 1997, 2006). The concept of contingency emphasizes how these assemblages develop and extend, not as the implementation of predetermined structures or the execution of detailed plans but, borrowing from Deleuze and Guattari's concept, as rhizomes that develop organically and capriciously (Deleuze and Guattari, 1987). Market assemblages do not move around a clear center that is located in a particular site; on the contrary, they are decentered wholes that may or may not be related to other assemblages.

De Landa's materialist and relational ontology omits the relevance of affects for constituting and sustaining assemblages, as well as agents' feelings of disaffection when these markets grow and become formalized. In this work, we contend that affects and other emotions, together with embodied values and lived experience, are the fuel of territorialization processes.

Methodology

Stemming from this theoretical perspective, we have researched the interface between practices of the members of a cooperative in southern Chile and processes that connect this experience to distant social sites that we identify as "global." This strategy allowed us to study the unique relational dynamics generated by agents involved in adapting and redefining food processes, policies, and public actions concerning responsible consumption. In doing so, this research focused on practices as the smallest unit of the social (Schatzki, 2002), and, by observing and recording such practices, we offer an analysis of the creative tensions that emerged during the process of territorialization of responsible consumption and the responsible consumer in Valdivia.

The case is based primarily on lived experience, which "seeks to enliven social theory by injecting living, breathing, feeling bodies into social methods and conceptual frameworks" (Carolan, 2011: 9). We, as consumers, members, and researchers, have been actively involved, through different roles and time spans, in the process of formation of the cooperative. We have bought products on a regular basis, participated in assemblies, sat in on board meetings, and joined social activities from 2009 onwards. Most importantly, we frequently cook and eat food from the cooperative. This allowed us, as engaged researchers, to experience the emotions and affects and to ethnographically reconstruct the processes, discourses, organizing practices and events that revolve around critical points of the case.

In the next section we present the case of La Manzana, in which we describe the heterogeneous associations of people, materials, and places that compose a new collective subject: the responsible consumer. Our aim is to present how "responsible" markets are assembled, focusing on the relevance of affects in the hard work involved in the constitution of collaborative relations and the emergence of subjectivities. To that purpose, we organize the case in three subsections: (1) friendship and enrollment, (2) crisis and commitment, and (3) growth and detachment. Following the presentation, we come back to the concept of food assemblage as a metaphor to understand the organic entanglement of human and other-than-human agents with crisis, contingency, affects, materials and place in the territorialization of responsible markets and in the emergence of the responsible consumer. In the last section, we offer some concluding remarks.

"Other consumption is possible": the case of La Manzana

Setting the place

Valdivia is a mid-size city, applying Chilean standards, with approximately 150,000 inhabitants located in the region of Los Ríos, 850 kilometers south of the capital, Santiago. The city is nested in the Valdivian Ecoregion, considered as a biodiversity hotspot. With a navigable river and surrounded by forests and wetlands, Valdivia's economy is based on tourism, agriculture, and forestry. Within the region, 34 percent of the population lives in rural areas. The Andes and coastal mountains population is predominantly Mapuche, the largest indigenous group in Chile, while the central valley includes descendants of German migrants. The main agricultural activities are horticulture and dairy and small-scale cattle ranching. In the coastal zone there are artisanal fisheries that provide the city's restaurants and the main river market with fresh seafood products. Of increasing importance for the region are the berry plantations and the hatcheries that supply the smolts (young fish) for salmon farming whose products are mainly exported. The city's so-called "knowledge economy" plays also a significant role: Valdivia has four universities, the largest with more than 14,000 students and 3,000 employees. Valdivia is also known for its active civic environmental networks and the presence of important international nature conservation NGOs.

Friendship and enrollment

La Manzana started in 2009, around a group of friends who were interested in buying healthy food directly from producers. Although Valdivia offers a number of large supermarkets, a year-round river market and a few weekly fairs – whose sellers are mostly middlemen – it is often unknown where the food comes from and how and by whom it is produced. As a result of agricultural modernization programs, in Chile most producers – big and small – use large amounts of synthetic fertilizers and biocides to increase production. In addition, much of the food sold in supermarkets is processed, canned or frozen, flavored, and contains preservatives, thereby changing the taste, smell, texture, and properties of fresh ingredients. Searching for products grown without biocides and processed without many chemical and unhealthy additives, the group of friends started to share tips about local producers in the area who were selling products whose quality or production methods were known. Soon, they moved from simply sharing information to jointly purchasing small quantities of different products. One of them would buy for those interested; afterwards, the product could be picked up at his or her place.

Many members of this initial group were not Valdivians. They had come to the city to study, work, or because they liked the city's lifestyle, generally characterized by Chileans as "green" and "human-scaled." The group also included foreigners who had settled in Valdivia and contributed with ideas and experiences witnessed in Europe and elsewhere. In fact, when the initial spontaneous purchasing group dissolved, it was a Spanish woman, Ines, who enthusiastically proposed starting a cooperative. Several meetings were held with a core group of approximately 30 families to discuss the project.

A number of issues had to be resolved during these first meetings. Chilean law prescribes that cooperatives cannot be composed of both consumers and producers and that consumers' cooperatives must have a minimum of 100 people to be officially registered with the Ministry of Economy. Hence, the first challenge was to enroll more people in the initiative. The 30 families began a charm offensive with friends, colleagues, and acquaintances to interest them in the idea of buying food from local producers.

Furthermore, the cooperative needed to make decisions regarding the products it would offer. The idea was to provide local products as a healthy alternative to those sold in supermarkets and local fairs and as a way of contributing to local development. Nevertheless, this raised questions such as: Would they include *only* products of local producers or also products from other regions in Chile? How would they decide on the healthiness of products? Should products be certified as "organic"? This last issue brought to light the fact that many small producers, who cultivate their crops without chemical inputs, cannot afford expensive certification procedures.

The cooperative decided to develop its own assessment method in order to decide whether a farmer was producing organically or was in transition towards agroecology. The cooperative also considered the farm's location. This was an important issue because in 2004, Valdivia and its surroundings suffered a huge environmental disaster caused by one of the largest cellulose factories in the world. This disaster caused deaths of large numbers of swans and other species in one of the largest wetlands surrounding the city. People were worried that the contamination had affected not only the wetland around the factory, but also the groundwater and therefore food production in the area.

Besides fruit and vegetables, the members wanted to include homemade products such as marmalade, sauces, cheese, butter, bread and cookies. Indeed, many people in the region still make those products at home, either to consume or to sell at local fairs. Due to legal constraints, however, the cooperative could sell only products coming from producers who were registered in the tax office and had an approved sanitary license to produce and sell processed food. In its early days, the cooperative had only three producers of fresh fruits and vegetables and a small number of authorized homemade products.

Although it was conceived as an organization with a horizontal structure, the cooperative required, in compliance with Chilean cooperative laws, a manager and a board. A member who had completed a course on cooperatives was proposed as the manager and appointed by consensus. It was decided that he would be in charge of daily affairs such as determining the supplies, filling in orders, paying the producers, managing the finances, and keeping the books. The cooperative would pay him a part-time salary. The board were volunteers. Upon its creation, each member of the cooperative paid an incorporation fee of 30,000 Chilean pesos (approx. US$ 50). These resources were destined to fund the cooperative's registration fee and to support its activities at this early stage.

In November 2009, the cooperative was formally founded during its first official assembly. At this initial stage, La Manzana functioned mainly with an order system. Members received a digital Excel file listing the products and its suppliers. These files were sent back to the manager, who calculated the number of products, contacted the suppliers, and organized their reception and distribution by volunteers. For that, the cooperative needed a physical space where the products bought could be stored and later picked up by the buyers. But they could not afford to rent a commercial place in the city. One of the members who lived in an old Valdivian house close to the center offered to sublet two rooms at a very low price. This place became the distribution center of the cooperative and a central meeting point for its members.

Crisis and commitment

During the first semester of 2010, the cooperative seemed to be working well. The members received, monthly at first, and weekly later, a list of products. They were able to pick up their orders twice a week: Saturday mornings and Monday afternoons. As members requested adding new products to the list, the number of suppliers increased. The cooperative decided, furthermore, on a strategy change: keeping certain items in stock. The distribution center turned into a little shop where a small number of products were permanently available.

Besides acting as an intermediary between buyers and sellers, the cooperative also organized social activities for its members and their families. The board decided to arrange farm visits as means of bringing producers and consumers together to talk about, prepare, and eat food. These interactions helped producers to value their products and learn customers' opinions, and for members to appreciate the work that goes into producing these goods. Furthermore, these activities aimed at creating and strengthening friendship ties between members. These visits and food sharing meetings were important for creating an identity, but the truth is that only a few members participated in these activities. Many of them had full time jobs, active social lives, and families; they did not have or make time to participate.

During the second semester of 2010, however, producers began to complain to board members that they had not been paid for delivered goods. Board members first thought that these were isolated incidents related to an organization in its initial stages. But later that year they realized it was a larger problem that could be solved only by changing the way the cooperative was working. The cooperative's operation was based on a projected sales volume that was much larger than actual sales. Furthermore, the cooperative aimed to pay producers a fair price and made little profit on each sale. As a result, it had been unable to pay all the suppliers' bills, the store rental, and, eventually, the manager's salary. The manager had kept silent about the financial problems. Since the cooperative was based on the principles of friendship and mutual trust, the board only came to recognize this problem when the initial capital had disappeared and, when a debt of 1.7 times that amount had accumulated. Total sales in 2010 rose to US$ 18,000, but after deducting operational costs there was a negative balance of US$ 8,000. The board decided to take control of management and to end the contract with the manager, who was also a friend/member.

The cooperative held an annual general assembly in April of 2011, focusing on the financial crisis and the question of whether to end or to revive the cooperative.

Approximately 30 people came to the meeting, all expressing their interest in keeping the cooperative alive. As the board received its mandate from the members, it proposed different options to solve the financial crisis. But instead of discussing solutions, people bitterly expressed their surprise about the fact that suppliers had not been paid. Given that the main principles of the cooperative were creating local and trustworthy networks between consumers and producers and engaging in responsible consumption, members were embarrassed by the situation. Some members also pointed to the lack of internal solidarity and commitment regarding voluntary work. Many families had joined the cooperative but were not actively buying, therefore, not helping to increase sales and revenues. In the next months the board decided to continue without the manager and, instead, looked for someone willing to accept this position without remuneration. The financial crisis implied a crack in the relations of solidarity both among the members and between the consumers and some of the producers.

The board organized a second extraordinary assembly to discuss how to handle the cooperative's financial debt. One of the options proposed was to ask a loan from a bank, just as some members had suggested when the cooperative was first created. Members present in the meeting rejected this proposal because they did not want to rely on financial institutions and preferred to get out of debt slowly by improving operations.

To tackle the most urgent debts, active members paid in advance for products they would receive later. The ordering and delivery system was optimized so that in-store stock could be reduced. It was also agreed that operational costs had to be drastically reduced by increasing voluntary work. This implied that board members became more involved in the daily activities of the cooperative, including working in the shop. In addition, thanks to the networks of its members, the cooperative started to receive foreign students – mainly from Europe and other Latin American countries, who were looking for volunteer opportunities. The survival of the cooperative, however, was possible due largely to the commitment of the suppliers. A significant number of producers accepted the board's proposal to be repaid gradually and, most importantly, did not stop delivering products so the cooperative could keep operating. Yet, others ceased delivering products until debts were paid and some withdrew from collaborating with the cooperative. Suppliers that continued working with the cooperative became more and more committed to the organization, the people, and the ideals of healthy fair food and local development.

During the second half of 2011, the cooperative slowly settled its debts. Although winter sales were very low, due partly to the limited range of fresh products available, the cooperative now had a low cost structure and could survive until spring when sales would grow again.

Growth and detachment

During the first two years, the shop was oriented primarily to members. But to settle debts and become financially viable, the cooperative needed to increase sales.

It was decided to offer services to the general public. To differentiate between customers, members received a five percent discount on their purchases. In 2012, sales were slowly increasing and debts with suppliers were settled. The yearly balance, however, was still negative. The board decided that boosting sales required increasing the variety of in-stock products available at the shop. For the convenience of members, the cooperative now began selling imported food products: goods such as cacao, coffee, and frozen fruit puree, often imported from international fair trade networks. The cooperative also offered imported organic or biodegradable cleaning and personal care products, such as diapers, shampoo, soap, female cups, and nutraceuticals, which were not available in the Valdivian market.

In 2012, the cooperative organized the First Meeting of Responsible Consumption and Fair Trade in Valdivia, a three-day event open to consumers, producers, cooks, academics, activists, and students. Seminars, debates, artistic shows, and gastronomy contests were organized. Local farmers and other producers offered their products in small stands. The success of this first meeting, based entirely on the volunteer work of members and their families, encouraged the group to repeat the event in 2013 and 2014. The meetings grew in size; more local actors helped to host the event and a larger public visited the fair and participated in the activities. Interestingly, the event began receiving sponsorship from the regional government and various state agencies that had previously refused to support the cooperative because the project was considered "too alternative." The Fourth Meeting, "Hands on Earth: agroecology on the table," was held in November 2015, and was organized by the cooperative with the participation of different organizations, authorities, and local and regional media. Due to its success, the initiative is currently considered as a benchmark for national cooperatives.

In December 2014, the cooperative received a major grant awarded by the Innovation Fund for Competitiveness of the Regional Government and the Foundation for Agrarian Innovation (FIA), an agency of the Chilean Ministry of Agriculture. After an open and competitive call, the board obtained approximately US$ 200,000 to carry out an 18-month project entitled "E-commerce and innovations in the value chain of agroecology of Los Ríos." Soon thereafter, the national undersecretary of economy visited La Manzana and gave a press conference at the store where she stated the importance of cooperativism for local economies.

A team of cooperative members developed various activities focused on strengthening relationships with suppliers, particularly with a group of 21 peasants from the neighboring commune of Paillaco. The fund also financed the redesigning and refurbishing of the store, the making of an electronic inventory and online sales system, renewal of the website, and the expansion of the use of social networks. In addition, new packages of agroecological and local products were designed and marketed through three channels: the store, web sales and home delivery. The main innovation was the agroecological basket: a weekly assortment of fresh, local, seasonal products that was delivered "from farm to door" using a van purchased for that purpose.

Beginning in 2015, the store has been open to members and the general public from Monday to Saturday. The bulk of the work is still voluntary, but the assembly decided to pay a small salary to the two main shop clerks. Sales have increased as new customers were coming to the shop. However, this has also created new demands. Some clients complained on La Manzana's Facebook page about bad shopping experiences caused by the "lack of professionalism of the personnel." Not knowing the history of the cooperative or its operation, they demand the shop to function as any other store.

During the sixth assembly, held in April 2016, the board announced an increase in sales to about US$ 70,000, which meant having a positive balance for the first time since 2010. The members applauded and celebrated this achievement with a collective open-air family lunch.

Extending the food assemblages: an open-ended experience

La Manzana can be understood as a rhizome extending organically through different surfaces, materials, and places, due to the force of affects of heterogeneous agents. These agents channeled flows of matter-energy – initially food and later other products – between producers and consumers and between different sites that were previously disconnected, and, in the process, forming partial assemblages of solidarity and/or reciprocity through the moment of consumption. The case shows that forming and sustaining such food assemblage required the enrolment of components from the state and conventional capitalist markets. For some agents, the process of growing and extending the assemblage triggered disaffection and instigated different responses: abandonment – such as the case of some members – or expressing anger and frustration as documented by some non-member customers who demanded from this shop the same standards as other commercial services. Nevertheless, as evidenced in the sixth assembly of April 2016, the prevailing feeling is that of genuine commitment to nurture an initiative that brings people together through responsibly produced items.

For the cooperative to survive, members had to overcome their dichotomous thinking and include elements that were not "purely alternative." When the cooperative first started, its most committed members avoided shopping for food at "conventional" capitalist food markets. Their aim was to change the "food system" through direct action. They also distrusted the state as an effective agent of social change regarding food matters, especially since the Chilean government has promoted agrifood policies based on economic clusters of export-oriented products. When La Manzana was at a critical point of bifurcation, however, its survival and thriving implied the creation of new markets. The cooperative required partial inclusion of modern managerial practices, learned from regular business and the use of different state instruments, such as applying for public funding, to support their ideas. Modern managerial practices helped the cooperative to organize a system that better embodied the principle of financial responsibility, key to sustaining trust and civic friendship between producers and consumers. On the other hand, the enrolment of the state through the FIA project provided substantial resources, which helped the cooperative to grow and strengthen work relationships with producers. For the state agencies involved, becoming part of this assemblage provided access to a concrete experience of a new phenomenon in Chile: organized civic movements promoting organic food and responsible consumption. The alliance between the state and the cooperative was widely publicized and reflects a discursive change in the country regarding rural and local development in which cooperativism regained significance in the construction of markets of solidarity and proximity. The inclusion of these elements by the board was contingent and pragmatic rather than planned; nevertheless, they were necessary to sustain the food assemblage.

We can see here how the process of territorialization of the cooperative is composed of heterogeneous elements, assembled by agents through care and affection to materialize principles. The dichotomy alternative/conventional would not be able to grasp this heterogeneity without excluding, *a priori*, what seems contradictory. The same applies to the global/local distinction, which loses meaning when we find that the cooperative's shop has become a hub for people, food products and ideas from around the globe. The global (i.e., imported products, volunteers from abroad, and foreign members) is embodied, re-embedded and transformed, just as the local (producers, products, and sites) reach the global through exchange and virtual existence. Here we have alternate movements of territorialization and deterritorialization operating without the assemblage being threatened, but mutating into new relationships and forms of organization.

We think the case of the Cooperative of Responsible Consumption La Manzana is relevant as a record of a civic food experience in southern Chile, a social site distant from the Global North and from the national center for decision-making. This case is also valuable as a theoretical reflection on the importance of affects in the assembly of new food markets. La Manzana evidences that changes in patterns of food consumption depend not only on structural productive conditions or on behavioral adjustment between supply and demand, as suggested by neoclassical economic theory. Nor are such changes exclusively the result of processes of cultural differentiation by upper- and middle-class consumers contesting conventional capitalist markets through rational choice. In Chile, where food quality is not necessarily a given as it is in Europe or North America, movements such as La Manzana provide not only new forms of quality control, but also the redefinition of food quality itself through direct relations between consumers and producers. Furthermore, the establishment of food solidarity has been an important goal for the movement in order to achieve concrete contributions to local development. As shown by the case, agents involved in developing La Manzana carried out actions, either planned or contingent, that creatively, but through hard work, instantiated new markets for consumption based primarily on their own search for local, healthy and fair food, thus creating new assemblages.

At the beginning of the chapter we argued that AFN approaches were insufficient to understand and explain new food movements as a result of at least three shortcomings. Through the use of assemblage thinking and an empirical detailed account of the case, we have demonstrated the inadequacy of narratives of totalities (the idea of a single alternative food geography made of different interconnected parts, and a universal food experience), the deterministic effects of dichotomies, and the importance of affects in the repositioning of food movements. We can affirm that the situated constitution of food assemblages is an effect of heterogeneity, multiplicity, and contingency, and therefore, each assemblage is unique. This does not rule out potential connectedness between assemblages of food solidarity. On the contrary, it provides grounds to reject totalities such as a singular food geography or a food world system (whether conventional or alternative). Instead, as a result of territorialization, we can distinguish many food assemblages, each continuously creating and transforming its own objects and subjects. Thinking through assemblages implies an active overcoming of dichotomies and the inclusion of a wider range of agents, affects, and experiences in the production and consumption of food.

As an end note, we claim that food assemblages, such as La Manzana, not only have a performative role of resistance, as some authors of the AFN's literature contend, but also an intrinsic value of intensifying existence, as they thrive on creative and novel associations between heterogeneous elements, opening up new ways of imagining, producing, and experiencing food related worlds.

Acknowledgments

This research has been partly supported by the project: FONDECYT N°1160857.

References

Anderson, B. and McFarlane, C. (2011). Assemblage and geography. *Area*, *43*(2), 124–127.

Anderson, C. R., McLachlan, S. M., McDonald, W. and Gardiner, J. (2014). Navigating the fault lines in civic food networks. *Journal of Agriculture, Food Systems and Community Development*, *4*(3), 79–99.

Arce, A. and Marsden, T. K. (1993). The social construction of international food: a new research agenda. *Economic Geography*, *69*(3), 293–311.

Arce, A., Sherwood, S., and Paredes, M. (2014). Repositioning food sovereignty: between Ecuadorean nationalist and cosmopolitan politics. In Trauger, A. (ed.), *Food Sovereignty in Geographical Context: Discourse, Politics and Practice in Place*. London: Routledge, 125–142.

Aubry, C. and Kebir, L. (2013). Shortening food supply chains: a means for maintaining agriculture close to urban areas? The case of the French metropolitan area of Paris. *Food Policy*, *41*, 85–93.

Brunori, G., Rossi, A. and Guidi, F. (2012). On the new social relations around and beyond food: analysing consumers' role and action in Gruppi di Acquisto Solidale (Solidarity Purchasing Groups). *Sociologia Ruralis*, *52*(1), 1–30.

Carolan, M. (2011). *Embodied Food Politics*. Burlington, VT: Ashgate.

Carolan, M. (2015). Affective sustainable landscapes and care ecologies: getting a real feel for alternative food communities. *Sustainability Science*, *10*(2), 317–329.

Coelho de Souza, J. Z. (2008). Comércio solidário na prática: o Núcleo Litoral Solidário da

Rede Ecovida de Agroecologia. MSc thesis, Desenvolvimento Rural, Faculdade de Ciências Econômicas, Universidade Federal do Rio Grande do Sul, Brazil.

De Landa, M. (1997). *A Thousand Years of Nonlinear History*. New York: Zone Books.

De Landa, M. (2006). *A New Philosophy of Society: Assemblage Theory and Social Complexity*. London: Continuum.

Deleuze, G. and Guattari, F. (1987). *A Thousand Plateaus*. Minneapolis, MN: University of Minnesota Press.

Goodman, D. (2004). Rural Europe redux? Reflections on alternative agro-food networks and paradigm change. *Sociologia ruralis*, *44*(1), 3–16.

Goodman, D., DuPuis, E. M. and Goodman, M. K. (2012). *Alternative Food Networks: Knowledge, Practice, and Politics*. London: Routledge.

Harris, E. (2009). Neoliberal subjectivities or a politics of the possible? Reading for difference in alternative food networks. *Area*, *41*(1), 55–63.

Ilbery, B. and Maye, D. (2005). Food supply chains and sustainability: evidence from specialist food producers in the Scottish/English borders. *Land Use Policy*, *22*(4), 331–344.

Lamine, C. (2005). Settling shared uncertainties: local partnerships between producers and consumers. *Sociologia Ruralis*, *45*(4), 324–345.

Marsden, T. and Arce, A. (1995). Constructing quality: emerging food networks in the rural transition. *Environment and Planning A*, *27*(8), 1261–1279.

Maye, D. and Kirwan, J. (2010). Alternative food networks: sociology of agriculture and food. *Sociopedia Isa*. Available at: www.sagepub.net/isa/resources/pdf/AlternativeFood Networks.pdf

Parkins, W. and Craig, G. (2009). Culture and the politics of alternative food networks. *Food, Culture & Society*, *12*(1), 77–103.

Ploeg, J. D. van der and Renting, H. (2004). Behind the "redux": a rejoinder to David Goodman. *Sociologia Ruralis*, *44*(2), 234–242.

Preiss, P.V. and Marques, F. C. (2015). Tendências no movimento de re-localização alimentar brasileiro: uma análise de Iniciativas Colaborativas de Compras. *Tessituras*, *3*(2), 269–300.

Renting, H., Marsden, T. K. and Banks, J. (2003). Understanding alternative food networks: exploring the role of short food supply chains in rural development. *Environment and Planning A*, *35*(3), 393–411.

Renting, H., Schermer, M. and Rossi, A. (2012). Building food democracy: exploring civic food networks and newly emerging forms of food citizenship. *International Journal of Sociology of Agriculture & Food*, *19*(3), 289–307.

Scarabelot, M. (2012). Construção de cadeias agro-alimentares curtas e papel dos atores em Nova Veneza, SC. MSc thesis, Desenvolvimento Rural, Universidade Federal do Rio Grande do Sul, Brazil.

Schatzki, T. R. (2002). *The Site of the Social: A Philosophical Account of the Constitution of Social Life and Change*. University Park, PA: Pennsylvania State University Press.

Sonnino, R., Torres, C. L. and Schneider, S. (2014). Reflexive governance for food security: the example of school feeding in Brazil. *Journal of Rural Studies*, *36*, 1–12.

Whatmore, S. and Thorne, L. (1997). Nourishing networks: alternative geographies of food. In Goodman, D. and M. Watts (eds), *Globalizing Food: Agrarian Questions and Global Restructuring*. London: Routledge, 287–304.

Winter, M. (2003). Embeddedness, the new food economy and defensive localism. *Journal of Rural Studies*, *19*(1), 23–32.

14

250,000 FAMILIES CAMPAIGN

The existence of flavor and taste

Stephen Sherwood, Ana Deaconu, and Myriam Paredes

Introduction

Gilles Deleuze's (2003) provocative, if controversial book, *The Logic of Sensation*, celebrates the raw, emotionally charged work of the Irish-born painter, Francis Bacon. An appreciation for Bacon's work led Deleuze to a sharp criticism of representation in art, but also of representation in broader social life (ibid.: 87):

> We are besieged by photographs that are illustrations, by newspapers that are narrations, by cinema-images, by television-images. There are psychic clichés just as there are physical clichés – ready-made perceptions, memories, phantasms. There is a very important experience here for the painter: a whole category of things that could be termed "clichés" already fills the canvas, before the beginning.

Deleuze argued that an artist's greatest struggle was to overcome the "figure" and to avoid the obvious shapes and contours of the object and capture the underlying primal "accident" of sensation.

Here, we join Deleuze in seeking to depart from a highly representational perspective of reality in food studies where we are confronted with a dominant and popular figuration: food movements as "resistance." The academic literature and activist movements alike summarize food activism as a "fight" that involves historically marginalized peoples (especially, the indigenous and peasant farmers) and their enlightened, sometimes self-assigned representatives against the avarice of global capitalism in industry, the state, and science (see for example, Petras and Veltmeyer, 2011). While we do not doubt that this imagery represents a particular aspect of food movements, resistance thinking is strongly founded in an abstract understanding of social change as primarily an issue of struggle and the occupation of formalized institutions.

We find that the resistance orientation in Latin America dramatically obscures certain actualities of food activism through the blinding clarity of a series of little questioned dichotomies: global–local, North–South, rich–poor, and consumer–producer. As summarized in the introduction to this book, in viewing people, their families, and social networks as dependent on or somehow not complicit in institutions, resistance ultimately portrays development as disembodied and people as merely victims of mysteriously disconnected, out-of-control forces.

For us, food activism, while not necessarily independent of resistance, is also about a dynamic that is far more emergent and visceral and therefore rooted in Deleuze's sensation. As summarized in Arce *et al.* (2015), social change in food is highly nuanced, contingent, and dependent on people's encounters and ability to open up new possibilities. The realities of food movements in Ecuador, as elsewhere, are increasingly cosmopolitan and cosmopolitical – that is, circumscribed not by gerrymandering or supply and demand, but rather through embodiments. Similarly, activism is largely a socially messy experience where one's boundaries are not simply determined by gender, race, nationality, class or title, but also through food practice.

The image of social movements as chiefly reactionary or defensive does not capture the emotion or energy that we, as activists, have undergone in and through food movements. For us, social change is arguably less about *fighting against* an essentially abstract and unknowable "global industrial food system," than it is about *fighting for* highly pragmatic interests, such as certain desired ways of planting, procuring, cooking, or eating.

In this chapter, we draw on Deleuze's criticism of representation in art in hopes of overcoming the modern "cliché" of resistance. In its place, we wish to capture the "sensation" of food movements by shedding light on the pre-representational moment of Deleuze's "accident." Thus, we adopt the view that food is part of an endless process of socio-material embodiment that is best explained through people's feelings, practices, and performances. We summarize this perspective through the metaphor of "existence," and we argue that existence underlays the potential of food movements to overcome the distracting imaginary that the desired future is largely a question of "fixing" institutions of science, the state, and the market.

In so doing, we examine two influential actors in the food movement scene in Ecuador that have become champions of social change in agriculture and food primarily through means of existence: (1) the *Colectivo Agroecológico* (from now on the Colectivo), an informal collection of some 400 individuals and collectives dedicated to alternative, "non-violent" means of agriculture and food, (2) the *Movimiento de Economía Social y Solidaria del Ecuador* (MESSE), a network of largely urban-based activists engaged in food circulation, creative marketing, and alternative currencies. These networks include a lively assortment of food activists working in seed saving, urban-based consumer wholesale markets, progressive gastronomy, start-up food fairs, and business enterprises as well as individual farms and urban households. In particular, we draw on the activity of the Colectivo and MESSE's ongoing "250,000 Families Campaign" (hereafter the Campaign).

The Campaign's sensation: flavor and taste

In 2014, the Colectivo and MESSE ambitiously proposed to recruit a critical mass of five percent of Ecuadorean households, roughly 250,000 families, to a self-determined and organized agenda of "responsible consumption" (Sherwood *et al.*, 2013). In concept, the Campaign is open to anyone who eats and is loosely led through self-initiative and democratic appropriation. Proponents of the Campaign include people from all classes and creeds, such as housewives, school children, farmers, urban "foodies," chefs, and development practitioners of every province in Ecuador. Since the Campaign does not organize by nation-state, it also has a number of cross-border initiatives, such as the ones in Colombia presented in this chapter.

While consumption is understood as the moment in which something is procured, for the Campaign, responsible consumption is an endlessly ambitious, vague and highly contestable concept. For some, responsible consumption may emphasize direct or non-currency purchasing, locally produced food, organic or agroecological production, or the use of traditional Andean crops and gastronomy (Jacobs, 2016). Understanding its objective as identifying and linking people around common or complementary practices, rather than seeking consensus, the Colectivo has been strategic in keeping the Campaign conceptually open and subject to debate.

Rather than seek predetermined learning through highly designed and orchestrated capacity-building, the Campaign employs open-ended, independent discovery, as espoused by John Dewey (1938). In particular, the Campaign seeks to recruit new practitioners through experiential learning in two fundamental entry points: flavor and taste.[1] Following Mouritsen (2015), *flavor* is understood as a sensation that encompasses the physical, chemical, and neurophysiological aspects of food. According to Deleuze (2003: 42), sensation has an immediate status in experience and indicates a pre-conscious form of being in direct contact with the world. As such, sensation is the level of "pure presence" (and potentially absence) and is conceived as the condition for sensational differences, such as vision, touch, hearing, and smell. Deleuze explains that while operating at the molecular level, sensations such as flavor do not merely question the relevance of scale, but suggest a difference in "kind." Or as he asserts (ibid.: 45), "Sensation is not qualitative and qualified, but has only an intensive reality, which no longer determines within itself representative elements, but allotropic variations."

Meanwhile, we refer to Gronow (1997) in viewing *taste* as an "aesthetic sociology" (in contrast to a sociology of aesthetics), a relatively robust, social concept that encompasses tradition, geography, and culture and, as per the thesis of Deleuze (2003), potentially human–nonhuman relations. In contrast to flavor, we view taste as a more multimodal and multifaceted concept, which may include how people come to perceive, value, and identify with certain food practices.

While the 250,000 Families Campaign is "faithful," in the sense of sharing an aspiration to unite people and their common interest in health, equity, and sustainability, if not survivability, the Campaign is something that cannot be heavily pre-designed or predetermined. In order to take social hold, responsibility, however practiced, must somehow become both primal and "tasteful." It must, through the biochemical vibrations of flavor and social commitment of taste, become a socio-material "fact."

Methodology

Drawing on Deleuze and Guattari's (1987) pragmatic approach to embodiment, here we are not interested in food as an object to be commoditized or governed in isolation from bodies, but as embodied. Deleuze and Guattari define a "body" as a sensorial, affective, and intensive set of associations that necessarily work both within and across a spectrum of scales, including a single organ and a human body as well as institutions of politics and law. As such, a body is not merely biological, material or social but simultaneously each. Assembled bodies are both different and interdependent, and they both define and are defined by circumstance.

An embodied approach to food studies demands study of people's reality as it unfolds. As per Deleuze's (2003) philosophy of art, painting the sensation is equivalent to recording the fact. It necessarily takes into account relationships as they are encountered and felt, rather than how one might imagine or wish them to be, thereby collapsing *a priori* representations of human experience and social organization.

We present three ethnographic accounts that are based on Deaconu and Sherwood's participation in the Campaign. When necessary, we change names of individuals to provide anonymity. We do not address the Campaign as it is publicly proposed or designed (e.g., its particular goals, objectives or ambition of recruiting the 250,000 families), but rather we study how it was experienced during three events over 2015–2016. As such, we place qualitative attention on its real-time activity, as forged through myriad emotions, intentions, contingencies, and flair. In particular we study: (1) the Madre Tierra Festival, held in the village of Tarqui in southern Ecuador between 2–4 October 2015, involving some 1,500 people from over 20 countries, (2) the *Aula Viva* (in Spanish, "Living Classroom") in the Amazon on the border of Ecuador and Colombia (19–20 November, 2015, involving 25 community representatives and organizers from Colombia, Ecuador, Peru, and Brazil), and (3) a "Healthy Food" workshop that took place at *La Cosmopolitana* Farm in the Rio Meta Watershed, Colombia (2–5 February 2016, involving 45 community organizers). As participants, we present these events in first person narrative.

Three Campaign events

Madre Tierra: regeneration of the planet

The Madre Tierra Festival, which took place in October 2015 and was organized by Ecuador's lively Seed Savers Network, brought together 1,500 individuals from over 20 countries for three days. The festival was grounded in 43 workshops on subjects as diverse as solar ovens, the history of agriculture, alternative currency, composting, bioconstruction, and more. Participants came independently or as representatives of a diversity of groupings, including community-based organizations, the provincial government, Slow Food, MESSE, the Colectivo, and many others. Based on Deaconu's fieldnotes, this account illustrates a typical Campaign encounter.

The interactive sensorial workshop on responsible consumption from the Campaign displays different fruits and vegetables in boxes. The participants feel the products with their hands and identify them through touch. Two volunteers from an audience of about 100 people – an indigenous Kichwa woman and a European-looking child with an Ecuadorian accent – are the first to enter the task. They squeal and giggle when they cannot identify what is inside. Upon pulling out the object – an achiote, they sing out with a collective laugh and move on to the next module, which involves the challenge of having to identify items without the aid of sight or touch, but rather just the tongue.

Next, volunteers interact with an oversized "nutritional plate," brimming with locally available grains, legumes, fruits, vegetables and herbs as well as instant noodles, cookie packages, a bag of household monosodium glutamate, and artificially flavored yogurt. The group is asked to identify what products they wanted to eat, and they put the others in a trash bin. The volunteers unanimously agree to toss the processed foods, but one teenage girl is left unsettled. She begins to gather the white rice from the plate with the intention of throwing it out. Met with surprise from the audience and the other volunteers, the young woman explains, "Here in Ecuador, we eat a Chimborazo [Ecuador's highest mountain] of rice for breakfast, a Chimborazo of rice for lunch, and a Chimborazo of rice for dinner."

The girl goes on to explain that this amount of white rice leads to obesity and diabetes and that the problem is not rice itself, but the poor nutritional quality of white rice and the high quantity consumed. She picks up a grain of brown rice from the plate and explains the difference: "This here, this is real food, and it tastes good. This we should eat." The activity continues into a tangential conversation in which the teenager fields the audience's questions on the different taste and texture of brown rice, where to buy it, and how long it takes to prepare.

The volunteers then select particular products with which they feel identified from an assorted variety of foodstuffs on a large cut-out nutritional plate. A highland Kichwa woman gravitates towards the different colors of corn and explains how each has a particular importance for her community. A Belgian man living in Ecuador and his preteen daughter focus on quinoa, highlighting its nutritional qualities and good taste. When asked before the audience about her personal favorite, the daughter looks befuddled, before speaking into the microphone with conviction, "What choice? We eat quinoa."

The final workshop activity involves a "sugar display": an assortment of about 12 commercial beverage products from the local store, including juices, sodas, and power drinks. These are exhibited with a small bag full of a measured amount of table sugar. The facilitator explains that each bag of sugar represents the precise amount of sweetener found in each bottle of soft drink, as per the indications on its label. Volunteers are called on to act out the expected reaction of a particular character to the sugar display. The first man chooses to be an incredulous politician, who is not willing to give up his beverage of choice, even fully knowing that it is "bad" for him. He explains that the immediacy of good flavor in the moment outweighs the potential menace of negative impacts in a far away future, jokingly referring to "beyond my immediate political term." Acting in opposition to the politician, a woman adopts the role of a concerned parent that is surprised to see the amount of sugar contained in a vitamin-fortified fruit drink that she regularly purchases for her child. "The advertising told me that this was healthy … I want my money back!"

After the workshop, a crowd from the audience eagerly approaches the sugar display to examine it in further detail. One man holds up in amazement a bag full of 42 grams of sugar – found in a 300 ml bottle of Coca Cola. An elderly Ecuadorean woman says in disbelief, "Is this amount really true?" Then, she asks advice on how to read and interpret the sugar identified in the label. An Argentine man jumps in with an explanation, while an oversized businessman says, "I no longer drink sweet beverages, because I have diabetes. I used to bring a two liter soda bottle to work every day, in lieu of lunch." This leads to a collective call to stop drinking sodas, and the group discusses a number of alternative beverages, such as fresh fruit juice, lemongrass tea sweetened with honey, and *agua de panela* (water sweetened with raw cane sugar). One woman declares, "These taste better anyway. I really don't even know why we drink soda."

Aula Viva: transforming dangerous food

Following the Madre Tierra Festival, Michelle O. Fried, an Ecuadorean nutritionist, popular cook and cookbook author, and member of the Colectivo, was invited to provide an introduction to the 250,000 Families Campaign for the Aula Viva – an Amazonian initiative, organized with the Pan-Amazonian Ecclesial Network (REDPAM) and the Provincial Jesuits Conference of Latin America (CPAL), with the purpose of "transforming the growing violence in the region into a force for peace and abundance."[2] Michelle is an elderly nutritionist who, while born in the United States, has lived in Ecuador for over 40 years. Through her books, Michelle has become an influential actor in the Colectivo and a champion of its 250,000 Families Campaign. Michelle worked with other members of the Campaign to hold a half-day experiential course on flavor and taste. Based on Sherwood's field notes, we provide an account of a field visit on both sides of the San Miguel River, dividing Ecuador and Colombia.

Sandra, a community organizer from the Catholic Church's Pastoral Social and co-organizer of the Aula Viva, picks us up outside our hotel in the Ecuadorean oil town of Lago Agrio, so that we can join the other participants of the Aula Viva. She explains that the group comes from different parts of the Amazon, but it is not made up of just ordinary people. Many are "masters" in managing the traditional Amazonian *chakra* – a form of agroforestry and permaculture that emphasizes multistory sunlight/shade management and the cultivation of multipurpose crops, trees, and even animals. "It basically resembles a jungle," she says, as she points out a densely populated, highly diverse plot of trees and bushes alongside the roadway. She adds, "We think of this agroforestry system as an edible forest."

The group meets at the homestead of a Colombian born couple that arrived in rural Lago Agrio some eight years earlier. Standing alongside his wife and surrounded by a collection of ragged looking *campo* dogs, the husband explains that after a violent incident in their community, he decided it was time to leave. In coming to Ecuador, he explains, "We did not want any gifts or giveaways. We came in search of peace and the opportunity to work the land with dignity." Recently, the couple sought help from the Pastoral Social to obtain a land title and to "technify" their farm, which according to Sandra, involves "the installation of a traditional chakra, a transition from annual to perennial crops and the introduction of multipurpose trees."

Following a walk around their 12 hectares of largely barren land, Martha, a colleague from the FUCAI who works with the Sikuani people of the Orinoco, had previously asked each visitor to arrive with a favorite Amazonian plant for donation. We form a circle around a large collection of seeds and plants of different colors, shapes, and sizes. There are different types of palm, what look like mahogany as well as a citrus, cacao, and a number of exotic fruit trees. With the visitors bowing their heads and cupping their hands and pointing towards the offering, Martha conducts an impromptu ceremony directed at the couple, during which she regularly pauses for the visitors to repeat after her:

> I have arrived with a seed from the hearth of my home to honor you for receiving me today. This seed comes from a plant that my family and I nurture as a vital part of our chakra. In return, each plant harvests the sunlight and provides my family energy in the form of the wood that cooks our food and heats our home; this plant provides shade, fresh air, and fruit. As this seed comes from a plant that provides nourishment for my family, I too bring nourishment to your chakra. This is my gift, our gift, of sustenance and life for you and your family.

When asked where best to plant the trees, Martha responds that there is no one "right way" to plant a chakra, before adding, "It's about connection and feel." A young indigenous man from Brazil adds, "A chakra is a living organism through which we simply flow." With that, the group uses sticks to trace a picture of the layout of the land on the ground, and the couple works with the group to identify where to place each species.

After planting the trees, we gather to say goodbye, so that we can move on to our next destination in nearby Colombia. The couple is noticeably touched, with the wife saying, "Our life is very hard, and sometimes all we think about is work.

We've never received such kindness." We drive a brief stretch of difficult, pot-holed road, which carries us across the San Miguel River into Colombia's Putumayo. One hour later we arrive at the town of La Dorada, Parish of San Miguel, where we meet Father Arturo, a Jesuit, to start a workshop on traditional Amazonian cuisine, as per the histories of the participants.

At midday we took a break to have lunch at a local restaurant. As the soup was served, stories began to emerge about the history of San Miguel, in particular with regard to the danger of fishing in the rivers and farming. Over the last two decades, the region had been a site of unspeakable violence, with all kinds of stories intermingled with the travails of gaining access to food during the civil war. According to the church representative, "So many people were murdered that we stopped keeping records. I only had direct knowledge of the names and cause of death of people who had received a proper Catholic burial. In my case, it's easy: the leading cause of death is murder, be it by paramilitary, the FARC, or the state."

As a consequence of the forced general strikes, thousands of families living in the otherwise lush and abundant forest of the lower Putumayo frequently face a crisis of food scarcity and hunger. One of the participants, a leader from a village near La Dorada, explains that for farmers and indigenous groups, the daily search for food has become a dangerous game of chance. "Going out to hunt or farm means facing the demands and abuses of the armed groups – both formal and informal. We face death through combat, fights, kidnappings, extortion or plain murder." She goes on to explain how even breathing in the Amazon has come to mean sickness:

> Some places in our department are strategic for hunting animals or for farmers to get their products to market. But those are also strategic points for the militaries. Movement on the river was restricted. Bridges were destroyed. Then the airplanes started to spray the coca plants with the poisons, though they usually missed and only hit our trees and crops. If you don't get shot, then working to produce food means getting sick. Every child on this side of the river has burns and sores from the chemicals. Our eyes burn, and we all suffer from the breathing sickness.

La Cosmopolitana: "flavor lasts a lifetime"

As a result of his experiences with the Campaign in Madre Tierra and at the workshop in the Putumayo, Roberto Rodríguez, founder of *La Cosmopolitana*, a natural reserve and learning farm located in the Meta River watershed, on the Orinoco, Colombia and a massive upland catchment area for the Amazon, decided to bring the experiential approach closer to home – to the families working with his foundation. In February 2016, he invited Michelle to lead a set of workshops under the title of "healthy eating." Taking the example from the Campaign, he expressed the goal of the workshops to promote healthy eating in a broad sense, involving nutritional health, health of producers, of the environment, and of local economies. The following is based on Deaconu's notes.

The Orinoco region is hot and humid. During an introductory walk upon our arrival, a guide stopped the group at the base of a large taro plant, locally known as *bore* and appreciated for its value as pig feed. Michelle, who has a broad knowledge of edible plants, said, "How wonderful to find you growing taro!" She spoke of its renowned nutritional value in the South Pacific, especially for the protein content of its young leaves. One of the staff members reflected, "I can't believe that we've been using such a nutritious plant to feed the pigs and not ourselves!"

The next morning the remainder of the workshop participants arrives. They include La Cosmopolitana's core staff, volunteer promoters from the surrounding valley and mountainsides and their family members as well as farmers, housewives, a few doctors, a linguist, a woman displaced from the Putumayo region, and others. In all, we are 45 participants. The workshop activities bring into focus different relationships with food – encounters of the nose and the tongue, childhood memories, ceremonies, understandings of the statement, "what is good for you" and each person's food history, taking into account the "body," family, and community.

Michelle provided an account of a recent trip to see her daughter in Senegal, where she had been a Peace Corps volunteer 40 years earlier:

> Flavor is something that finds the soul and lasts a lifetime. … I remember when I walked off the airplane onto the tarmac in Senegal, and I smelled a familiar peanut sauce. I unexpectedly and uncontrollably broke into tears. Over four decades had passed [since I was in Senegal], and I had never forgotten those intimate sensations that so marked an expansive moment of my youth and life!

Each evening, Michelle leads the group in cooking a meal together using local ingredients that the participants bring with them or that grow on the La Cosmopolitana farm. One person pulls out a *ñame*, a fibrous yam that is commonly eaten in coastal Colombia and grows easily in this region, but whose culinary uses are largely unknown. In the Orinoco, most farmers grow ñame only for animal food. Michelle sought the group's suggestions for how to prepare the food. Mireya, an outgoing school teacher, observed, "It looks, feels, and almost tastes like a potato; ñame would be a good for a mashed-potato-style dish." She then volunteers her recipe for an onion-cheese sauce to accompany it. The flavor is delicate and delightful, and everyone digs in for more until the bowl is left clean. While many express their joy in being able to finally *eat* a crop that they have been producing just for sale, Ninfa, a leader from Las Leonas community, finds the taste of this unexpected recipe so agreeable that she is going to procure ñame for the workshop to be held later in the week in her community.

Although it has not been planned as part of the workshop, circumstances on the farm have brought up the aspect of multi-species in food and "energies" when raising an animal. The workshop facilitators and participants are captivated by a sow that has just given birth to a litter of 18 piglets. Having limited experience with pigs, many individuals from the group gather around the mother each morning and each night, marveling at the size of the litter, at how cute the piglets are, and at how much food the animals will become when they grow up. Jokingly, people name the piglets according to the different pork preparations that they would someday become. On the last day of the workshop, the sow dies, and La Cosmopolitana staff are left with the interminable task of bottle-feeding the piglets every two hours. When asked why she died, the caretaker explains matter-of-factly, "Because sometimes, when there are a lot of people around after they give birth, the sow gets stressed. We say it has something to do with the energies and *espanto* [an illness coming from intense fear or from negative energy]."

Discussion: fresh sensations

> It is not I who attempts to escape from the body, it is the body that attempts to escape from itself by means of ... in short, a spasm.
>
> *(Deleuze 2003: 15)*

Connected to different expressions of unfolding life (and death) in food, each Campaign experience follows its own pathway. In the case of the Madre Tierra Festival, the 250,000 Families Campaign worked with an open-ended sensorial experience to explore taste and flavor, hoping to bring new energy to *existence*. This happened through a public exploration of the range of flavors and pleasures in food as well as a discussion on food alternatives and the use of sweetener in purchased drinks. In San Miguel, embodiments of food sensations had come to include the different effects and affects of long-standing civil war, petroleum extraction, coca production and drug trafficking as well as broadscale aerial spraying of herbicides. In part, this inspired a movement to "rescue" the chakra and promote agroforestry and permaculture and a pursuit of "abundance." At La Cosmopolitana the "healthy eating" workshop explored traditional foods and surrounding alternatives, the production of flavor through recipes as well as a brief, but perhaps significant interspecies encounter between humans and pigs. Each of the three events utilized flavor and taste to spark different degrees of energy, fear, and reflection as well as hopes and aspirations for continued or new embodiments.

Deleuze (2003) describes sensation as a form of "spasm" – namely, the vibration or rhythm involved in creative discovery, and as such, a movement in the body, not between bodies. Accordingly, flavor is a chaotic, bodily materiality, before the body becomes ordered through intentionality and movement, such as the spasm that brought tears to Michelle upon arriving in Senegal and rediscovering the smell of a long "forgotten" peanut sauce. As another example, we find the visceral fear of looking for food in the war zone of San Miguel. The spasm, while seemingly singular, can also be residual. Rather than being part of a global narrative on what a return to West Africa or living in the Amazon should be, existence is rooted in the particularities of doings and their residual materialities – a coupling between sensation and life's embodiments that leads to what Deleuze (2003: 67) calls "involuntary memory." As experienced in the discovery sensorial workshop, the local food explorations or the pursuit of permaculture, this experience was potentially a sensation of lasting consequence.

The Campaign as well as its partners at FUCAI and the La Cosmopolitana had made a social commitment to the notion that human–food encounters were capable of generating endless ways to sense, eat or, farm as well as to live and be. During workshops there was no discernible prescription of what an ideal meal should contain or an edible forest should look like, opening up space for experimentation. Finding herself among opposing parties of violent conflict in the Amazon, instead of analyzing the histories, the FUCAI leader opted for the creation of an edible forest, thereby creating the possibility for a new existence.

Arguably, the promotion of sensation through human–food encounters was not enough in and of itself to explain social change around food. It was flavor linked to the socio-material organization in the practice of taste that led to new associations and the materialization of food embodiments. These embodiments did not begin with a blank canvas. Fields, animals, and food products are full of a residual history or memory in which the cook, the eater, or the farmer assembles and is assembled and, as a "painter," paints – even before a sensation is fully understood. The spasmic sensation of flavor and the material act of taste are central to bringing the canvas to life.

Conclusion: "Faith in life"

> Bacon tries to eliminate the sensational … when he said, "I wanted to paint the scream more than the horror."
>
> *(Deleuze 2003: 28)*

The experience of the Campaign aligns with Deleuze and Guattari for whom being depends on human–material relations that embrace context, material, and people and involves an assemblage of extensive–intensive capacities. Change in food is born in the constitution of embodiments, enabled through common (and not-so-common) practice – what we summarize through the metaphor of existence. The boundaries of space are not drawn just by legal documents or the scriptures of a formalized institution, but more so, they are inscribed through the "memory" of practice. In this sense, we argue that the Campaign's influence is largely bound by the socio-material and biological potentials of flavor and taste.

In its work, the Campaign foments sensation around food that is not merely experienced as a rational explanation of health, environment or economy, but rather it is also lived through laughter, surprise, sadness and delight. Food becomes a common site for social commitments – sometimes among people and objects that otherwise might not find common ground – for unique co-creations of a "tastier" world, beginning with their own moments of consumption and bodies. In the Aula Viva, flavor and taste have created a space for identity making, which in turn reshapes and displaces coexisting or competing senses of place, bounded not merely by geography or institutions, but also experience and practice. Ultimately, flavor becomes a transformative moment and taste a transformative space, working in solution and precipitating in practice seeming binaries of conception/action, theory/practice, mind/body.

Deleuze (2003: 67) said, "There are two violences: that of spectacle and that of sensation. The first has to be renounced to reach the second. It is a kind of declaration of faith in life." People who had experienced different forms of disembodiment, as expressed at the Madre Tierra, the Aula Viva, or La Cosmopolitana, were not unaware of the history of environmental degradation, inequity, or cruelty and hurt in their family and neighborhood. Instead, they chose to "destroy" one clarity with a new clarity: the reproduction of a flavor, the recuperation of the chakra, or the pursuit of abundance. One could say that in addressing the different forms of violence associated with food, the Campaign's initial response was, "Savor this!"

In escaping the temptation to narrate a single representation of the multiplicities of food, the Campaign seeks to supplant it with primal flavor and emergent associations and embodiments such as a new dish, a new fruit, agroforestry and permaculture, or an emergent form of organization, encounter, or exchange. The Campaign focused on the primal energy of food that sometimes is embedded in stories of violence, rather than in the violence itself. The participants utilized this energy for new purposes, namely a plethora of sensations and other potentialities tied to (re)constituting "faith in life."

Acknowledgments

The research reported here was part of a Dutch-financed NWO–WOTRO Programme on "Catalyzing Positive Deviance in Agriculture and Food" and Canada's International Development Research Centre research project on "Ecuador's Healthy Food Campaign." The opinions expressed here do not necessarily represent those of the donors or their respective Board of Directors.

Notes

1 One good example of the Campaign's open-ended, discovery-learning approach to learning is: Keely McCaskie (ed.) (2013). *Nutrición y alimentación en la Sierra Norte del Ecuador: guía para facilitadores comunitarios*. Quito, Ecuador: Fundación EkoRural.

2 The founder of La Cosmopolitana, Roberto Rodríguez, defines abundance as the energy and life sustained through the Amazon's unique material and biological "wealth," with specific reference to the region's genetic and biological diversity as well as its rich natural resource base (titanium, gold, uranium, cobalt, bauxite, magnesium, silver, diamonds, water) and fossil fuel reserves (oil). He argues, "Abundance is not something that one can acquire, but rather it is something with which one must connect" (Field notes, 19 November 2015).

References

Arce, A., Sherwood, S., and Paredes, M. (2015). Repositioning food sovereignty: between Ecuadorian nationalist and cosmopolitan politics. In: A. Trauger (ed.), *Food Sovereignty in Geographical Context: Discourse, Politics and Practice in Place*. London and New York: Routledge Studies in Food, Society and Environment, 125–142.

Deleuze, G. (2003). *Francis Bacon: The Logic of Sensation*. Translated by D.W. Smith. New York: Continuum.

Deleuze, G. and Guattari, F. (1987). *A Thousand Plateaux: Capitalism and Schizophrenia*, Vol. 2. Minneapolis, MN: University of Minnesota Press.

Dewey, J. (1938). *Experience and Education*. New York: Collier Books.

Gronow, J. (1997). *The Sociology of Taste*. London and New York: Routledge.

Jacobs, L. (2016). *Responsible consumption as a vehicle for social change in Ecuador: a study of translation over controversies in food*. MSc thesis, Knowledge, Technology and Innovation, Wageningen University, the Netherlands.

Mouritsen, O.G. (2015). The science of taste. *Flavour*, 4: 18.

Petras, J. and Veltmeyer, H. (2011). *Social Movements in Latin America: Neoliberalism and Resistance*. New York: Palgrave Macmillan.

Sherwood, S., Arce, A., Berti, P., Borja, R., Oyarzun, P., and Bekkering, E. (2013). Tackling the new materialities: modern food and counter-movements in Ecuador. *Food Policy*, 41, 1–10.

15
CONCLUSION

The vitality of everyday food

Stephen Sherwood, Eleanor Fisher, and Alberto Arce

In this book on social change and development in agriculture and food in Latin America, we have sought to address how agriculture and food unfold within the vitality of everyday experience. Born from an ambition to further our conceptual understanding of how food matters to the organization of people, the book can be read as an ethnographic critique of a romantic vision of norms and the aggregate capacity of institutions. In contrast to such abstractions, the chapters draw on highly empirical research to illustrate how people play a crucial role in creating real-life events. By fulfilling democratic values, they provide relevant instances capable of moving food debates beyond the present-day controversies over food security and sovereignty.

From a methodological standpoint, the chapters share a commitment to real-time study of food practice and a "flat" ontological perspective – that is, view free from assumptions of *a priori* relationships between people and food based, for example, on gender, race, or class. While such social categories may be useful for describing differentiation, their application can become self-reifying, leading to little questioned suppositions on continuity and change.

In their study of a range of ongoing struggles, negotiations and accommodations involving different actors, the authors explored how people draw upon different repertoires of knowledge, assets, values, and forms of civic movement to create new forms of corporeality. The resulting book is regionally situated, politically engaged, and conceptually emboldened. Interested in strengthening the empirical grounding of food studies, we view this as part of an ongoing reappraisal of the relationships embodied in food production, circulation, and consumption.

In continuation, we explore what this collection contributes to our collective insight into agriculture and food developments in Latin America. At the end, we propose a brief research agenda for a more critical approach in food studies.

Agrofood realities: challenging common wisdom

As summarized in the introduction to this book, the agrofood studies literature is rife with accounts of the revolutionary fervor of peasant farmers and urban-based consumers organizing to fight against the avarice of an out-of-control "global food system." In contrast, the empirical work presented here – from the bio-intensive Tenate food basket in Oaxaca, Mexico (Chapter 10), water for food production (Chapter 5) or the use of milking technology (Chapter 6) in Ecuador to the New Dawn Milk Cooperative in Mar del Plata, Argentina (Chapter 12) – identifies another dynamic: one characterized not just by *a fight against* a system or regime, but rather *a fight for* highly pragmatic materializations of values. The case material shows that food is not merely constructive (i.e., it is not solely the product of intention or design); food is also born and borne through a nexus of relational encounters. As such, food is not just the intentional expression of identity, but it also is a product of the relations generated in and through interactive processes of production, circulation, and consumption. In other words, food is both constructed and constituted through situated practice.

Social studies in agriculture and food, particularly as applied in socio-environmental studies, are often based on essentialism and positivism, especially when representing worlds that are epistemologically defined as a natural state. This normative pathway has led to a preoccupation with a search for social principles and basic rights. Over time this perspective has become the central agenda for a proposal of state-based governance that seeks a predictable, stable society that is capable of managing contemporary risks. Nevertheless, as the chapters throughout this volume reveal, the social world is fundamentally open, unpredictable, and creative. The case material finds no underlying principles or universal explanations in the social world, but only processes, relationships, objects, and corporealities, as generated through daily interactions with food.

The chapters call attention to new relational forms enabled through food that emerge and catalyze actors, things, affects, and bodies into a new corporeality of existence. In so doing, the different experiences permit a grounded reflection on relational being and public good as forged through contemporary food practice – namely, saving seed and planting a field, bringing water to plants, milking a cow, delivering, procuring and preparing food, and weaving a hat as well as the endlessly visceral sensations tied to eating. Most basically, the resulting processes stimulate multiple, contested events and materialities that question – theoretically, methodologically and normatively – the long-standing social science tradition of separation between object and subject.

In the authors' examination of contemporary agrofood embodiments, we find that research and engagement with bodies – both public and civic – involve multiple and overlapping socio-political relations tied to the senses as well as to particular life histories. We find that the research provides the basis for better understanding how people and things bring forth coexisting food-worlds, as it announces a need for further analytical attention to common, and at times discrepant, agrofood materializations. The chapters present how experience is constituted through a multiplicity of materialities. In effect this work travels beyond a subject–object divide towards infinitely creative, if contingent, flows in which food is embodied and given corporeal expression.

Several chapters, but most notably Silva *et al.* (Chapter 11), adopt an ethnographic view of food as a public good, fundamental to people's social life and well-being. This particular study exposes a struggle between state-supported crop production solely for flour against a well-established cottage industry for artisanal bread and straw hats, leading to competing territorial configurations and assemblages. Other examples include a multinational celebration, consumer responsibility, and citizen-organized food consumption and political mobilization. Taken together, these cases reveal how food contributes to ways in which different bodies of knowledge are constructed and constituted.

The chapters highlight an increasing dissolution of spatial boundaries dividing jobs and processes (production, consumption, and eating), administrative units (farms, markets, government agencies, and cooperatives) and countries, generating ebbs and flows of information, images and emergent properties among individuals, social groups, and entities that coexist in a variety of modes of being, all of which demand food – as a public good in a multiplicity of forms, shapes, and quantities. The authors describe diverse interweavings of practice, values, and ideas that take material form, including a collective food basket, a common crop or meal, a milking machine, a paradise tourism destination, a formalized farmer's network, and an international cooking event. These embodiments and corporealities are sites of the re-assembling of relationships of (ill)health, (un)sustainability, and (in)equality in wealth and power.

The literature on food sovereignty commonly places the dominant actors of food pleasures and travails in a tidy narrative of oppositions involving marginalized, victimized peasant farmers and helpless consumers facing powerful abstract entities: transnational corporations, the state, science, and an out-of-control global food configuration. Likewise, in its effort to identify natural relations and to generalize agriculture and food as tied to a global transcendence of alternativeness, food studies literature often overlooks people's empirical reality – that is, the actualities of life at home and on the streets. The notion of a critical transcendence in food studies postulates that processes of food globalization can only be contained through state regulation, ignoring, for example, the gap between government discourses about sovereignty and an excluded people eking out a living in some remote territory. One can question whether national regulations can truly allow citizens to set their own agenda for inclusion and for ending political marginalization. In addition, one can ask which part of the nation-state can be trusted to support people's actions and their values of environmental sustainability, social justice, dignity, and fair income.

Nevertheless, our main preoccupation with this politically correct and transcendental analysis is more specific. We ask: what do nation-state policies mean when key institutional actors in the state, but also in social movements and global forums, distance themselves from people's reterritorialization of trade, labor, environment, health, and food? Food studies commonly loses sight of the contingency in food as well as the creativity of emergent embodiments and their corporealities and multiplicities (i.e., a manifold network of processes that interact with one another), which often lie outside cognitive categories or the expectations of how life uniquely unfolds. In this vein, we argue that the *immanence* (i.e., the realm of "God's presence within" or, in other words, the mundane) in food studies involves a democratic and inclusive politic of life, in which actors involved in the nation-state, social movements, and multinational corporations, among other spaces, seek to extend influence and control. The notion of an international public movement for equality and social and environmental justice may be difficult to achieve outside the description and empirical analysis that reveals the material existence of efforts to assemble globalization processes, be it through resistance from below or via existence from within. A number of chapters in this book, but in particular Oscar Reyna's work on the Wirikuta nomadism and rebellious Peyote (Chapter 7), analyze such efforts and reveal a potential public ontology that, while being more than merely social, challenges the popular institutional epistemology of regulations through which the vast majority of present-day food studies seeks to explain transformations in agriculture and food.

Meanwhile, the daily experience of family farmers in the Galapagos Islands (Chapter 8) demonstrates how organic, home-grown and locally marketed food can effectively block the leading source of invasive species to the islands (i.e., the introduction of fresh food from the continent) and decrease dependence on fossil fuels (due to dependence on food brought in by ship from the 1,000 kilometers of distance between the islands and the Ecuadorean mainland), reminding us that repetition in the shared understandings and values of environmentalism, but also in food studies, has come at the cost of seeing the creativity in the actualities of people's daily encounters with food. A social construction of paradise leads to notions of ecotourism in the Galapagos, indigeneity in the Amazon (Chapter 3), and producer–consumer relationships in places such as Oaxaca and Valdivia (Chapter 13). In the process of attempting to generalize human experience, social scientists commonly overlook nuanced opportunities, for example in nature conservation, the protection of a culture, or the capability of bio-intensive farming to unleash affordable, healthy food. The studies presented here diversely explain food as a public experience, transformation, and active embodiment with the potential of providing what Gilles Deleuze or Manuel de Landa (as explained in Chapters 12 and 13) might describe as rhizomatic access to food, environmental concerns, rights, and the politic of justice.

Leaky embodiments: finding the flow of existence

Bringing to the fore the matter of the body, the significance of food (dis)embodiments reinforces how issues of food quality raise new policy concerns for national governments, civic movements, growers, retailers and consumers as rapid economic, urban, and rural changes become manifest in people's physical-material being. Over time this has become reflected in, for example, national health statistics on morbidity and mortality rates. Beyond concerns over the immediate human body, health, and nutrition in relation to food, the cases in this book demonstrate how collective embodiments become a vehicle for political action and environmental (dis)content. Here we have seen how particular foodways incorporate (un)sustainable production and consumption practices, thereby influencing notions of place, power, and risk. Feeding into this are everyday experiences of (ir)responsible production, circulation, and consumption that enable or disable certain food associations.

Taken together, the chapters position the body at the center of emergent social relations between nature and culture, subject and object, and inequality and justice. With the breakdown of a series of oppositional categories (urban–rural, consumption–production, rich–poor, and North–South), the actors begin to generate new food corporealities, engendered through their practice. In analyzing this, the authors (re)problematize human and nonhuman actors in assemblages, thereby revealing the political significance of bodies in action. This activity reveals the multiple ways in which food is linked to experience and the consumption of modernized polluted environments, to affects (memory, fear and nostalgia), to the yearning for exotic foodstuffs, and to social policies that address basic public services. It also is tied to a proliferation of "public bads" – the unwanted, yet increasingly undeniable collective products of modern food, such as mass pesticide poisonings, pandemic overweight/obesity, and global warming.

Our distinctive public ontology focuses on the notion of multiple subjectivities, in which actor's bodies rhizomatically associate material things (such as a landscape or a food product) with nonhuman entities (culinary knowledge, bacteria, and microorganisms). Such human–nonhuman associations intensify the experiential qualities of situation and place (e.g., the taste of a particular wine or recipe), thereby generating space for action, creativity, affect, and common sense. The resulting embodiments become a potentiality in the hands of actors involved in public events, the production and distribution of goods, or the activities of competitive or cooperative coordination. In the process, new subjectivities may emerge that are capable of objectivizing (i.e., distributing the object of food in rhizomatical fashion through different subjects) emergent materialities, as observed in the nomadic use of Peyote in the Wirikuta Desert as well as during the Ajijic Chili Cook-offs in Mexico (Chapter 1). The authors explain the manifestations as the outcome of temporary, unstable, "fluid" materiality. These food corporealities – be they in the form of seamless human–Peyote relationships or a valued pot of chili – are what Michel Foucault calls "somatic singularities," which constitute a contemporary unknown within a horizon of imperceptibility and specific difference.

The empirical research in this book poses a much needed challenge to the orthodoxy of social science on agriculture and food in Latin America, in particular with regard to the largely unquestioned reliance on reference points that focus on state, science, and the market. By implication, environmental and social goods tend to dwell on rationality and formalized institutions. While a critical project does not abandon the expectations for a better future, it preoccupies itself with the practicalities of survival. As explained by Paola Silva and colleagues in Chapter 11, food can be considered as a public good, which for them implies a being that cannot be treated merely as natural object, but rather food also can be part of assemblages of distributed subjectivities, involving both the human and nonhuman.

Coming to the fore is how agriculture and food are the outcome of largely expressive, non-dialectic processes of being and becoming. The case material presented here – from bringing water to bear on food in Ambuqui and the appropriation of government-supported technology in Santo Domingo to the constitution of modern rural diets in Carchi and the flavor and taste of childhood food in the Putumayo – sheds light on the contingency, agency, and self-organization in agriculture and food as a configuration of assemblages from within and without, rather than against or from below. This perspective contrasts with approaches that emphasize how agrofood actors (and entities) perform according to universal principles or conditions or some other preordained schemata. In the terminology expressed in a number of experiences in the book, including the Tenate food baskets in Oaxaca and the 250,000 Families Campaign in Ecuador (Chapter 14), our empirical work from within food movements does not find a mere struggle of resistance, but rather situated and pragmatic endeavors of "momentum" and "sensation" found in *existence*. Accordingly, food is not simply a matter of understanding, representation, and authorship, but more so, it is a part of unpredictable and even inexplicable processes of formation, invention, and fabrication as well as accident.

Food engagements: public goods and their potentials

The public goods that emerge through assemblage and association in the different cases are organized and made effective by means of practice and voice, and they come to be through diverse *modus operandi* (procedures, techniques, and rituals) that effectively coordinate and restrain the imagination, quality, and possibility. In short, the public good is a singular association that comes into being as a state of politic – a special force that interfaces with the national state and the market. From a methodological perspective, the public can be utilized to observe practice and its results in order to explore the characteristics of such neglected resources as food lifeways and to account for the "facts" of daily experience.

As diversely encountered in this book, the public good exists regardless of one's willingness to recognize or acknowledge it. Not only are present-day institutions in Latin America increasingly blinded to particular public goods, but through a growth in hybrid public–private partnerships, they also have systematically neglected people's experiences and actualities. In an effort to defend the citizenry through protection of the commons (of access to land, water, flavor, nutrition, seeds and markets as well as the governance of knowledge), these institutions systematically decrease the space of the public. As we see in contemporary Latin America, this is taking place in the name of enhancing biosecurity, intellectual property rights and patent protection, food security, and food sovereignty.

The new conflict is not just between public and private, but also between people and the authority they have bequeathed to formal institutions – be it the state, science, or the market – to manage the public good, as embodied in knowledge, territory, and exchange. In the process, people today are not just in danger of losing their representation and recognition in society, but also at risk is their control over their food territories. As found in such diverse locations as Santa Domingo, the Galapagos, and Valdivia, food embodiments and their emergent corporealities diversely overcome this tendency through at least two relational processes: the collapse of dualities and dichotomies and de/reterritorialization.

Multiple chapters, but in particular the struggles of farmers in the Galapagos with the impositions of conservationists, the state, and the tourism industry, reveal how cultural images and objects, such as the notion of paradise, are constituted and objectified through the indeterminacy and flux of cultural life. In this way, lived experience and culture – emergent through the interplay of actors, knowledge and sites – become the focus of bodily being-in-the-world.

Being and becoming: actors, knowledge, and sites

Together, the chapters shed light on the means of being and becoming. In the cases presented here, an actor signifies one or more people in pursuit of individual or collective interests in a social network. Actors' actions are an important agency resource for new identities and new ways of life within the modernities of the contemporary world. It is here that the embodiment of politics centers on the multiple subjectivities of livelihoods, which may converge with issues of health, food quality, and food experience. As found in the purchasing groups in Oaxaca and Valdivia, the New Dawn Milk Cooperative in Mar del Plata, Argentina, and the *Articulação Nacional de Agroecologia* in Brazil (Chapter 9), networks are an agency resource of high potential. Socially situated and disembedded from institutional political realities, networks are embodied in actor alliances and performances and territorially located throughout food sites, where agents create space and may come to wield extraordinary degrees of civic political influence on contemporary institutional decision-making.

Knowledge encompasses not only cognitive knowledge but also the affects, experiences, and skills gained through life, stored in the physical landscape and built into the environment and social interactions. Knowledge is people's public understanding of their social categories, such as ideas, relationships, and affects of relatives, friends, neighbors, and institutions, but also their practice, such as in eating out. Practices of eating generate social and natural interactions in the streets, allowing people to observe one another's food-knowledge. As called for by Gross and colleagues in Chapter 2 and experienced throughout Latin America, further research is needed to visualize the continuities and discontinuities in existing diets, describing how specific social communities (e.g., schools or clubs), through everyday sayings and doings, generate new forms of knowledge, however harmful or healthy, in providing access to relatively inexpensive, high-sugar and high-fat foods. Such research provides insight into how political public influence hinges on the embodiment of human–nonhuman alliances – experience that could be particularly useful provided the pressing need to respond to a growing overweight/obesity pandemic in children, youth, and mothers through the mobilization of knowledge resources found in food (i.e., non-industrialized or non-processed food), people (their healthy living and being) and their social institutions (norms and values).

Sites form the territory of multiple interconnections or assemblages of human coexistence with the vibrancy of food. This may include, for instance, the physical (e.g., towns) and even imaginary sites (such as the virtual worlds of television and the internet) or those spaces where social practices happen in households, schools, and in the streets. Sites are spaces of food (re)connection with territories and culture. Here, global processes are mediated, managed, or attuned to the potentials of such food territorialization through the life experience of people.

The mobilization of bodies and food contributes to the awareness that internal and external relations should not be seen as creating different orders, but rather together they are involved in (re)constituting fragments and parts, which modify an actor's subjectivity and physiognomy – both human and nonhuman. Their organic bodies are rooted in natural and social interfaces, which in turn become significant metaphors, imaginaries, and imaginings. At these sites food territorializes itself through the strength of associations that are embodied in a variety of selves, symbols, and meanings.

Relational forces: intersubjectivity and political alliances

In connecting corporeality, affect, and reflexivity, processes of embodiment demonstrate how food practice, knowledge, and technology break through divisions, such as those that commonly separate object and subject or nature and culture. New relationships and alliances are formed at an interface where entities (real or otherwise) construct, reconstruct, and dismantle themselves in the co-constitution of individualities – be they linguistic or corporeal. Following the work of Deleuze and Guattari, a collection of chapters (6, 7, 11 and 12) refer to this process as uniquely intersubjective. Through intersubjectivity, the experience of creativity is available as an emergent resource not just to the individual, but also to the nonhuman other, thereby generating associations between individuality, the social, the self, and the other. We view such intersubjective processes as central to the birth of a new public good.

From an ethological perspective, territory has to do with the environment of a group – for example, the Andoque in the Colombian Amazon (Chapter 3), ecotourists or colonists in the Galapagos, agroecology networks in Brazil, or the New Dawn Milk Cooperative – that cannot be objectively located. Instead, territory is determined through processes of constitution, in this case as rooted in agrofood practice. As per Felix Guattari, *relative* deterritorialization retains the possibility of reterritorialization, such as domestication by the state, science or the market. It refers to transcendence, as commonly explained by present-day food studies. In contrast, *absolute* reterritorialization involves the impossibility of going back to an earlier state or territory, referring to the constant production of *immanence*.

While certain cases cite the dangers of people handing themselves over to the transcendence of science, the state, or the market (e.g., by indiscriminately accepting the recommendations of the *ingenieros* or conservationists, giving in to formal processes of certification or going back to the temptation of the supermarket), as explained by Gustavo Blanco *et al.* in Chapter 13, many of the experiences appear to be in processes of absolute deterritorialization in the form of immanent embodiments, corporealities, and multiplicities. Experiences such as the growth in agroecology networks or urban-based consumer movements are shaping new territory that does not necessarily obey the boundaries of the genetics of indigeneity, the economics of class, or scientific standards of "best practice."

Andrew Ofstehage's study of transnational soybean farmers from the United States in Brazil's Cerrado (Chapter 4) and their toils with contingency, improvisation, and unpredictability reveals the complexity of new agricultural and social relations. These are sometimes contradictory to what traditionally has been understood as the rural community. This raises questions over commonly held beliefs in the benefits of the modernization and globalization of food. It also highlights the extent and intensity of socio-environmental degradation. The isolation, alienation, and broadening of food production and consumption are a complex result of the assembly of endogenous and exogenous entities, which ultimately places into question the notion of their dichotomy.

It is from such relationships that we understand the cosmopolitanization of food and agriculture as a reflexive response to rising rates of mobility, fluidity, and interdependence between work and the expressions of agricultural goods, as presented in the Ajijic Chile Cook-off and the 250,000 Families Campaign. In part, this is due to the emergence of highly intensified experiences of people, acting as consumers in search of the original flavor and taste sensations of the familiar as well as the exotic, leading to all sorts of expected and unexpected outcomes, including goodwill, generosity, charity, and peace. These experiences illustrate that agricultural practices are not necessarily a part of a market, nor do they necessarily organize and operate in line with the "natural state" of a territory, legal institution, society, or culture.

In its legalization of private property and franchising of public property, increasingly the nation-state in Latin America is consuming the public, thereby melting away both individual and common property – as perhaps best illustrated by the controversy over wheat germplasm ownership and biodiversity in the Bio-Bio territory, Chile. In this case, public agriculture policy for crop breeding conflicts with local practice, organized around other purposes – namely, wheat for the preparation of bread and artisanal hats. Recent official recognition of farmers' "hidden" wheat varieties has begun to provide a degree of social legitimacy for long-standing territorial handicraft and culinary knowledge. Such socio-political-material experience reveals a multiplicity of cooperation, coordination, and intra-actions with certain imperceptible, random happenings, such as earthquakes and market events, shedding light on the intersubjectivity of everyday life, in this case the existence of a human–nonhuman agency as a potential quasi-reality involved in the territorialization of change. In the Bio-Bio case, food and handicraft have value not just because these are in popular demand, but because people and their territory derive value from the knowledge that other varieties of wheat for food and handicraft exist. The public of the territory maintains these goods because their existence and their consumption are intrinsic to the territorial commons. This experience suggests that such assembled relationships merit further attention as a means of better explaining agrofood relationships and their effects and affects. As such, public good can be seen as not just a concept, but also as a pragmatic and useful approach for working with people to identify and catalyze territorial potentials for social change and development.

Corporeality: relational being

In their study of the constitution of responsible consumption through the *Cooperativa La Manzana* in Valdivia, Blanco and colleagues challenge a series of little-questioned dichotomies: North/urban/wealthy/modern/bad and South/rural/poor/traditional/good. In addition to the problematic organizing of states of being as good or bad, in their preoccupation with the presence of the good, formal institutions can lose site of the "bad." In effect, such binaries blind the observer to the rich goings on in food. In the process, the living presence is overlooked and food is dispossessed of its bodies, and food studies fails in its responsibility to factually describe and make sense of food, but also to critically inform policy. In contrast, the authors find the centrality of affects in the process of creating a new subject (the responsible consumer) and a new object (the food cooperative and fair food), which leads them to call attention to a need to go beyond structural and materialist understandings of an assemblage.

The literature has created lists of representations for agroecology and food movements, and while there may be certain truth to these descriptions, there is always difference. The nature and potentialities of food are always elusive. For example, the chapters on changing patterns of nutrition in Carchi and the Chili Cook-off in Ajijic reveal how contemporary food becomes embodied within lived experience, thereby mobilizing minds, objects, and relationships in ways that exceed classic representations. Emergent food lifeways are simultaneously specific and dynamic, as migrating rural families or foreign retirees and Mexican nationals publicly build social relationships and live together in a locality. In comparison to the dichotomies that divide classes or nationalities, youth and the elderly, the relations forged through the constitution of food can even lie outside the imagination or even the possible.

Civic movements represent and promote alternative agendas for change, which under certain circumstances challenge forms of normative authority and orders that are seemingly dominant. Nevertheless, the expectations of people, with their family dreams and the effects of belonging to geographical or imagined communities, give rise to new forms of authority dedicated to mediating change. The resulting inhibitions, when met with those brought by the "external" expert, industrialist, or bureaucrat, generate actions that cannot be explained solely as deeds to materialize disobedience and resistance against the forces of domination and globalization. These forms of organization, in favor of the immediate existence and daily life, are more than just an expression of a rationality of social groups with political consciousness. This type of effect and policy as "course of action" causes the explanations that use the contradictions of the earlier mentioned categorical dichotomies to collapse and fall apart, revealing the importance of the description of everyday experiences that support life forms, existences, and manifestation that are empirically difficult to deny (although they clearly do not fall within the representations of the macro-influences or conceptual micro-frameworks). What is left is simply relational living and being, in all of its gloriously undeniable circumstance, contingency, and coincidence.

The vitality of everyday food

The contributors presented here challenge, from a range of standpoints, normative biases concerning how agriculture, food, and development *should* be. Instead, the authors place their analytical lens on how everyday experience unfolds through diverse modes of living and being – namely how agriculture and food *is*. What remains are the contingent effects and affects of the daily life, leading to the identification of multiple everyday food embodiments and assemblages constituted out of a willfulness that, we argue, is capable of surpassing hegemonic scientific wisdom and established neoliberal discourse. This leads us to propose a fresh perspective for conceptualizing subjectivity and agency in relation to food and its material forms.

While we would not discount the many valuable contributions to date of institutionalism and governance in agrofood studies, considering its neglect of the rich socio-biological and socio-material dynamics involved in the different embodiments, corporealities, and assemblages found here, we are left questioning the continued utility of mainstream critical food studies, such as the present-day preoccupation with certification. We argue that further theoretical, methodological and normative unpacking is called for.

A number of chapters call attention to the limitations of common dualistic and dichotomous categories in explaining continuity and change in agriculture and food. In other words, is the world still (if it ever were) organized and dominated by North–South, rich–poor and urban–rural? In light of the proliferation of unwanted developments in agriculture and food, is it fair and promising to continue to rely on governments, science, and the global capitalist system to define and coordinate the course of action for people?

We propose a revised critical perspective that is both more situated in the actualities of agriculture and food and more critical in its attention to how agriculture and food are uniquely embodied as well as its willingness to acknowledge and embrace emergent multiplicities, however (un)intended or (un)desirable. This work gives prominence to the endless potentials of peri-capitalization (as per A.L. Tsing, the specific manifestations of capital assets at the margins of existing worlds) to overcome the stagnation and even recession of seemingly insurmountable sources of influence and power. The social agency and creativity found in each of the chapters justifies further qualitative attention to the multiple goings on in informal institutional spaces and people's self-organization – be it in the form of a native people and its intimacy with plants and the environment, an emergent purchasing collective, the interfaces of producer associations with cows and milking machines or irrigation technology, or encounters between networks and social movements vis-à-vis science, industry and the state.

A number of insights begin to identify an agenda for a more genuinely critical treatment of change in agriculture and food. These include renewed scrutiny of the mundane immanence of daily living and being – in the family, neighborhood and social networks as well as in administrative bureaucracies of the state, market, industry, and science. In particular, notions of place and space must be reconsidered as being not only bound by the physical and geographic but also by the virtual and social. In addition, there is a need for rethinking the relations of time as not merely a matter of succession. Past and present also are a matter of instinct, memory, and coexistence, as illustrated by the Andoque utilization of gold and the Wirikuta consumption of Peyote as well as the cosmopolitan context of the ongoing social reconstruction through flavor and taste in the Putumayo. Such a treatment of time opens the way for further attention to the constitution of intersubjectivities – the emergent intuitions, resonances, and dissonances that can be felt through the different rhythms and vibrations of being and becoming in agriculture and food.

From a methodological point of view, a strengthened critical perspective requires renewed commitment to empirical study and in particular to *practice* as what Theodore Schatzki describes as the highly situated, yet infinitely ambiguous "site of the social." In addition, placing emphasis on a flat ontology demands careful thought to normative preconception and appreciation for the particular sites, spaces, and time of unfolding life. As highlighted by a number of the contributors, this ontology must include scrutiny to anthropocentrism and openness to human–nonhuman relationships. This only can be achieved through access to the real-time dynamics of inhabitation in daily living and being, which in turn requires full immersion in the field, participant observation, and assiduous ethnography.

Beyond the work of institutional studies that emphasizes state-based or project-based development, this empirical treatment shows how people in the street create space for change, if in ways that commonly lie beyond their control, guidance or intention. In a period of growing cynicism, skepticism, and despair over the prospect of a better future in agriculture and food, the contributions represent a testament to human willfulness, pragmatism, and practicality. While humans may not be in full control of their worlds, we find that they nevertheless have a will for power and production, mutual affection, and embodiment. Food practices and values are revealed through the expression of creativity and intention – a vitality that despite, and even because of, seemingly insurmountable forces and odds provides a modicum of hope and inspiration for more healthy, equitable, and sustainable futures.

INDEX